Life at Small Scale : Behaviour of
Microbes.

Life at
Small Scale

Life at Small Scale

The Behavior of Microbes

David B. Dusenbery

SCIENTIFIC
AMERICAN
LIBRARY

A division of HPHLP
New York

Acquisitions Editor: Jonathan Cobb
Development Editor: Susan Moran
Project Editor: Kate Ahr
Text and Cover Designer: Vicki Tomaselli
Cover Image by: M. J. Grimson and R. L. Blanton
Illustration Coordinator: Susan Wein
Illustrator: Fine Line Illustrations
Production Coordinator: Paul Rohloff
Composition: Progressive Information Technologies
Manufacturing: Quebecor Printing

Library of Congress Cataloging-in-Publication Data

Dusenbery, David B.
    Life at small scale : The behavior of microbes / David B. Dusenbery
        p.   cm.
    Includes bibliographical references and index.
    ISBN 0-7167-5060-0
    1. Microbiology.   I. Title.
QR41.2.D87   1996
578—dc20   96-28629
                                                            CIP

        ISSN 1040-3213

© 1996 by Scientific American Library

Printed in the United States of America

Scientific American Library
A division of HPHLP
New York

Distributed by W. H. Freeman and Company,
41 Madison Avenue, New York, New York 10010
Houndmills, Basingstoke RG21 6XS, England

First printing, 1996

This book is number 61 of a series.

To All Who Share in the Wonder of Life on Earth

Contents

Preface

Assuming that other people see the world the way we do sometimes leads to misunderstandings; it is even more risky to assume that other animals have the same knowledge about their world that humans have. It is not even safe to assume that animals have less knowledge than humans—dogs learn more about the smells of their world than we do and "lowly" insects see ultraviolet light. As we consider organisms more and more different from ourselves, we find it increasingly difficult to view the world from their perspective.

This challenge reaches its epitome when we consider microorganisms. Most people would doubt that microbes (or plants) know anything. But writing this in Atlanta, in April, with the dogwoods and azaleas in full bloom, it is clear that plants know the time of year very well. Similarly, careful observation and experimentation have revealed that all organisms—even bacteria—obtain information about their environment. My goal in writing *Life at Small Scale* has been to celebrate this fact and to describe what science presently knows about the capabilities of microbes.

Organisms reveal the information they have gleaned about their environment primarily by their behavior, because useful behaviors must be closely attuned to specific local conditions—so this book is generally about the behavior of microorganisms. It is these behaviors that allow pathogenic microbes to invade the human body, and other microbes to decompose dead organic matter or fix nitrogen,

recycling the nutrients that make life possible. In selecting both the kinds of organisms and the kinds of behavioral activities to include, I decided to cast a wide net and not allow myself to be limited by the usual definitions.

Animal behavior is often defined in terms of movement and muscle contraction—but only because movement is the activity most easily perceived by highly visual humans, who usually study animals animated by muscles. A more comprehensive view (based on function) is that behavior includes actions that change an organism's environment—either by moving the organism to a new location or by modifying the old one. Consequently, this book includes the release of chemicals as well as locomotion among the behaviors discussed, and the reader will see that releasing chemicals into the environment is a particularly valuable behavior for many microbes.

Microbiologists commonly study bacteria, fungi, protozoa, or algae. But what distinguishes these organisms from other organisms? I have adopted a functional definition—that microbes are organisms sufficiently small that they can prosper without a circulatory system. This way of defining a microbe leads me to include rotifers, nematodes, and flatworms along with the traditional microorganisms. On the size scale at which all these organisms live, many physical relationships are different from those humans are subject to. These differences in the physical world form yet another theme of this book. The reader will find that many of the bizarre behaviors of microbes—their methods of locomotion, manner of feeding, means of communicating—are adaptations to the different laws of physics that apply on the micro scale.

I would like to thank all my colleagues in the scientific enterprise who have painstakingly gathered the observations reported here. Especially important are those who have taken the time to photograph some of the wonders they have found. The artistic eye and stubborn detective work of Travis Amos made possible the inclusion of many such photos herein. I am also grateful to the Metolius Meadows editing group for their enthusiastic polishing of the original manuscript. Several expert reviewers—John Bonner, Ken Foster, Alexander N. Glazer, James L. Gould, and Jeffrey E. Segall—made important contributions. I owe a large debt to Susan Moran of the Scientific American Library, who helped make the organization and writing more accessible. And finally, I thank Sharon for the sacrifices she made so that I could complete this project.

David B. Dusenbery
April 1996

Life at
Small Scale

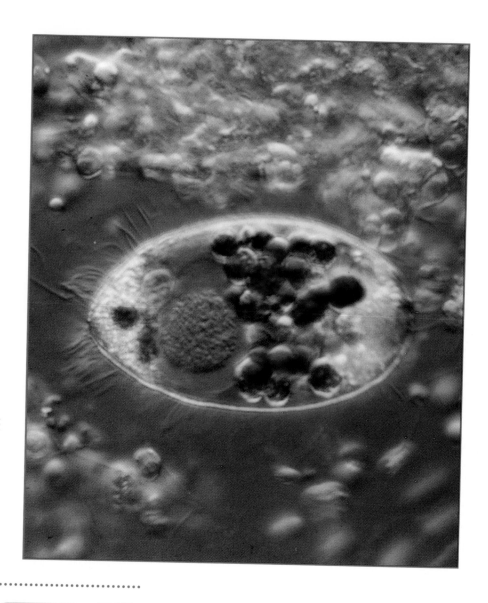

This protozoan, called *Orchitophyra stellarum,* lives symbiotically with a sea star, feeding on cells in the gonads. It moves by means of the cilia seen faintly around its edges.

Invisible Organisms

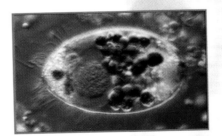

About five thousand years ago, archaeologists tell us, humans invented the wheel. Well, perhaps they did—but they did not invent rotary motion. A billion years before the first humans evolved, bacteria were swimming by rotating helical screws. Just like the wheels, pulleys, airplane propellers, and ship screws that humans now construct, the motors of bacteria have a rotating shaft and bearings. These structures are a million times smaller than the human-made devices, however, and it took scientists three hundred years of study to even begin to understand how they work.

In the summer of 1674, Antony van Leeuwenhoek collected a sample of cloudy water from a lake near his home in Holland and examined it through a microscope of his own construction. There he saw a whole new world of previously invisible organisms, a world so unsuspected that many people could not believe that the creatures inhabiting it existed. One of the most interesting features of the organisms that van Leeuwenhoek observed was that they moved and behaved in a manner that "was wonderful to see." Biologists have been intrigued by the behavior of the organisms inhabiting this new microscopic world ever since—and even more so since learning that these organisms act as chemical factories supplying the rest of life with oxygen and crucial nutrients.

Many of these organisms "obviously" moved under their own power. But this conclusion was cast in doubt in 1827 when the Scottish botanist Robert Brown observed tiny pollen grains moving as he examined them through an improved microscope. The announcement of his observations inspired a flood of theories and experiments aimed at determining the cause of the movement. The particles were first assumed to be alive, but every kind of small particle was found to exhibit similar motion. Scientists tried unsuccessfully to attribute these motions to gas formation, capillary forces, the illuminating light, convection cur-

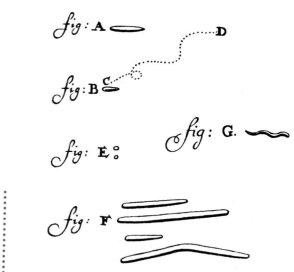

Leeuwenhoek made these drawings, published in 1683, to depict bacteria that he had observed through his microscope in a sample taken from the human mouth. He illustrates the classic forms of bacteria, and in some cases, considering the source, we can guess the actual species that he saw: A is a *Bacillus*, B is *Selenomonas sputigena* swimming in a spiral path, E are micrococci, F is *Leptothrix buccalis*, and G is a spirochete, probably *Spirochaeta buccalis*.

rents in the solution, and on and on. Nearly a century later, the calculations of Albert Einstein and experiments of the French physicist Jean Perrin finally solved the puzzle (and in so doing helped confirm the atomic model of matter). The explanation of "Brownian" motion was simply that any particle suspended in a liquid is constantly bombarded by molecules of the liquid. If the particle is small enough, the collisions will be sufficiently large and unevenly distributed to cause the particle to move erratically in a random pattern.

In spite of the confusion caused by Brownian motion, it remained clear that some bacteria moved much more rapidly and through larger distances than

Brownian motion alone could account for. These bacteria were apparently self-propelled, but the mechanism remained mysterious.

In the nineteenth century, scientists looking through the light microscope observed long tails, or "flagella," extending from many motile bacteria. Although the diameter of the individual filaments making up the flagella was well below the resolving power of any light microscope, many bacteria have several filaments that clump together to form a screwlike bundle, and these bundles of filaments could be seen by applying special staining techniques or darkfield illumination using bright sunlight.

In the first half of this century, scientists could see that the filaments were helical in shape when not moving and that, when moving slowly, they appeared to wave back and forth as expected for a helix rotating on its axis. Nonetheless, some scientists concluded

MICROCOSM dedicated to the London Water Companies. BROUGHT FORTH ALL MONSTROUS, ALL PRODIGIOUS THINGS, HYDRAS, AND GORGONS, AND CHIMERAS DIRE. Vide Milton.

MONSTER SOUP commonly called THAMES WATER, being a correct representation of that precious stuff doled out to us !!!

A view from the early nineteenth century of one reaction to the discovery of microscopic organisms. The artist presents some familiar and some fantastic organisms as present in London's contaminated Thames River.

that the flagella were not the cause of bacterial loco-motion. This uncertainty was finally settled in the 1950s when scientists sheared the filaments from the cell. Before shearing, 90 percent of cells had filaments and 90 percent were motile, whereas after sheering, 1 percent had filaments and about 1 percent were motile. It seemed that bacteria couldn't move without their filaments. Even after losing their filaments, the bacteria normally regained motility within a few minutes but would remain immobile if exposed to a drug that inhibited protein synthesis. Since the filaments were composed of pure protein, it seemed logical to think that the immobility was caused by their absence.

Even though the helix looked like it was rotating, few scientists seriously considered the idea that the filaments could rotate with respect to the cell, like a rotating shaft. As late as 1960, the major review article of the decade dismissed the idea "on the reasonable assumption that the base of the flagellar bundle is unable to rotate relative to the bacterial body." Instead, most researchers assumed that a helical shape propagated as a wave along the filament as a result of one part of the filament sliding or contracting with respect to adjacent parts. This is how muscles move, and everyone assumed the same mechanism worked for bacterial flagella—and it does turn out to be the mechanism used by the larger and more complex flagella of animal cells.

During the 1960s, however, several scientists began to propose that the filaments were rigid and actually rotated at the point where they attached to the cell. The first strong evidence for rotary motion was based on the realization that in a rotating bundle of helical filaments, the individual filaments must change relative position. Thus, a chemical that cross-linked one filament with another, would stop the rotation of all filaments. As predicted for rotary motion, when investigators applied antibodies directed against flagellar

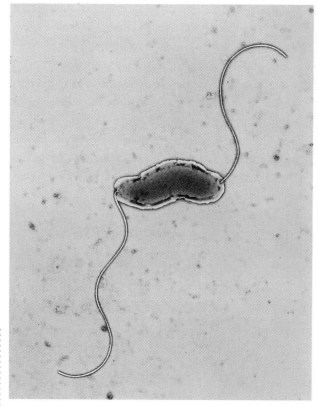

A colorized transmission electron micrograph (TEM) of the bacterium *Campylobacter jejuni,* which causes food poisoning in humans. These bacteria, curved or helical in shape, can swim rapidly by means of thin filaments that act like propellers. *C jejuni* is a common cause of gastroenteritis in people, especially children, who may acquire the infection from contaminated poultry, meat, and milk. Symptoms start with fever and acute abdominal pain, followed by watery and often bloody diarrhea. This cell is about 2 micrometers long.

protein, those antibodies able to cross-link flagellar protein molecules immobilized bacteria with many filaments but not bacteria with single filaments. The evidence became even more convincing in 1974, when Michael Silverman and Melvin Simon of the University of California at San Diego found a way to see the rotation. Taking cells from a mutant strain of bacteria that had only the stump of the filament, they used antibodies to attach the stumps to a microscope slide or to the stumps of other cells. Because the stump was stuck in place, whatever mechanism turned the filament caused the entire cell to rotate in the opposite direction. It could be seen that the cells rotated rapidly—several times a second. Since then a wide variety of evidence has accumulated supporting the notion that the filaments actually rotate like the shaft of a motor.

Flagellar rotation is but one example of microbes behaving in a way that is foreign to our common experience. As you continue to read, many more surprising behaviors will unfold, including mysterious gliding locomotion, microscopic catapults to disperse progeny, external digestion, the secretion of antibiotics, and sensing objects at a distance without the use of eyes or ears. In many cases these behaviors do not exist in our familiar world, and we will see that they are made possible by differences in the laws of physics at the microscopic scale compared to our familiar macro scale. So intriguing are these differences that physicists have become interested in exploring how and why microbes behave in the way they do. They have found, for example, that at the micro scale, viscosity is much more important than inertia in determining fluid motion; chemicals can be transported faster by diffusion than by the circulation of liquid; and the generation and detection of sound is impossible.

The Invisible World

Although we are usually unaware of their presence, the vast majority of all organisms are microbes. Generally speaking, the smaller the individual organism, the more individuals present in the environment. Thus, a handful of soil typically contains a few insects, thousands of tiny, wormlike nematodes, and millions of bacteria. It is possible to make a more subtle comparison by looking at the types of organisms. The first attempts at classification divided all organisms into plants and animals. But recent molecular analysis has revealed that there are more than a dozen groups of microbes that are as distinctively different at the molecular level and as distantly related as animals are from plants. These groups range from bacteria to the myriads of larger single-celled organisms, called protozoa, to yeasts and fungi. Thus, in spite of the much greater familiarity of large plants and animals, most organisms and most kinds of organisms are microbes.

Microbes live wherever large organisms live—in fact your body houses more bacterial cells than human cells—and they live many places large organisms cannot live. Even deep underground there are huge numbers of microbes. This surprising discovery was made only within the last decade, when viable bacteria were isolated from core samples taken hundreds of meters below the surface of the ground. Previously microbiologists had considered that living organisms were confined to the top few meters of soil. This may represent the discovery of a whole new biological world, independent of the organisms we have known.

As a result of their large numbers, widespread occurrence, and unique metabolic behaviors, microbes have a large impact on the earth's environment—from generating the oxygen in the atmosphere to recycling the nutrients of dead plants and animals. In short, we could not live without them. Understanding their behaviors may help us get along with them even better—bacteria causing serious disease often differ from harmless bacteria only in possessing a specific behavior that aids in invading the body or spreading from one person to another.

All organisms—microbes as well as plants and animals—must solve the same kinds of problems: how to obtain nutrients and energy, avoid being eaten, and disperse progeny, and how best to allocate resources between these activities. Yet because of the physical properties of the micro world, a microbe's solution to these problems may be quite different from a larger organism's. For an example of how these physical properties influence behavior, I turn again to locomotion: If rotary motion is so useful to bacteria as well as to humans, why do animals not use this wonderful device? To answer this question, it helps to understand something about how molecules spread throughout a space.

Moving Nutrients by Diffusion

Molecules are always in motion at a rate that is influenced by temperature, and the higher the temperature, the faster the motion. In fluids (gases and liquids), the molecules collide with one another and bounce apart like billiard balls. Under ordinary conditions, the average time between collisions for any molecule is very small (10^{-13} second in liquids and 10^{-10} second in air).

These repeated collisions cause individual molecules to move in an irregular path, similar to a random walk (described by mathematicians as a path generated by taking each step in a randomly chosen direction). A consequence of this random motion is that molecules tend to become evenly distributed. If a group of molecules—say, in a drop of perfume—start out close together at high concentration, but are free to move in a larger space such as a room, they will in time spread out and become evenly distributed throughout the space. This is the process we all know as diffusion.

During diffusion, molecules move, on average, from regions where they are present in high concentration toward regions where they are present in lower

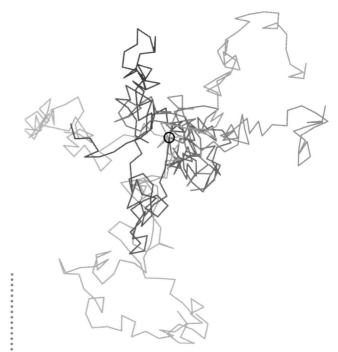

Six molecules start out close together (at a high concentration) and undergo a random walk, with each path represented in a different color. After a time, they have diffused apart, representing a lower concentration.

concentration. In the case of diffusion along a pipe, the net movement of molecules is proportional to the concentration gradient—the difference in concentration between the ends divided by the length of the pipe. The net movement is also proportional to a quantity called the diffusion coefficient, which depends on the frequency of collisions and the length of the path between the collisions. Diffusion is slowed by shorter path lengths from one collision to the next. For molecules diffusing in gases such as air, the path length between collisions is much longer than for molecules diffusing in liquids (where molecules are in continual contact). Consequently, diffusion in air is much faster than in water—about 10,000 times faster.

can provide a rate of transport sufficient to meet metabolic needs. But for transport over larger distances, energy must be expended to pump fluids from one place to another. Thus, our bodies must expend energy in breathing and in blood circulation, and the cessation of either is quickly followed by irreparable damage and death. In contrast, organisms that are much smaller in size can rely on diffusion to transport chemicals.

Let us consider more closely the size at which diffusion becomes inadequate for transport. As an example, we will use oxygen, a substance most cells need to extract energy from food. (Other nutrient molecules will diffuse at about the same rate.) An organism has to be able to provide oxygen throughout its interior. Since all organisms are mostly water, it is the rate of diffusion through water that limits the size of an organism without a transport system.

Assume that the model organism is spherical in shape, which is a good approximation for most single-celled microbes, and lives in pond water. Furthermore, assume that every hour the organism uses up on average 1 milliliter of oxygen per cubic centimeter of its body volume. (This rate is typical for small mammals, higher than the rate for most fish, and at the lower end of what is found for many bacteria.) Diffusion must bring oxygen into the cell fast enough to replenish the oxygen that is consumed and maintain a steady concentration. Assume further that oxygen is present

A distribution of 1000 particles after undergoing a random walk of 1000 steps starting at the point marked by the red cross. The particles represent the distribution of concentration after the release of a pulse of some chemical at the starting point. The concentration, or density of particles, is highest at the starting point and declines as one moves away. If the particles continued to follow a random walk, they would eventually be distributed evenly in the available space.

All organisms must provide for the transport of chemical nutrients into their bodies and the transport of waste products out. Over small distances, simple diffusion of chemicals from high concentrations to low

Diffusion through Air and Water

Medium	Mean free path length, μm	Mean collision frequency, collisions per second	Diffusion coefficient, cm²/s
Air	0.1	10^{10}	0.2
Water	0.00001	10^{13}	0.00001

in the surrounding pond water at a concentration in equilibrium with the atmosphere. As oxygen is consumed inside the organism, a concentration gradient will be established that ranges from high levels far from the organism to its lowest level in the center of the organism, and there will be a net flow of oxygen down this gradient from the environment to the interior of the organism. After sufficient time has passed, the concentration gradient will achieve a steady state in which the concentration is constant in time at any location. How will the concentration vary with distance from the center of the organism, and will suffi-

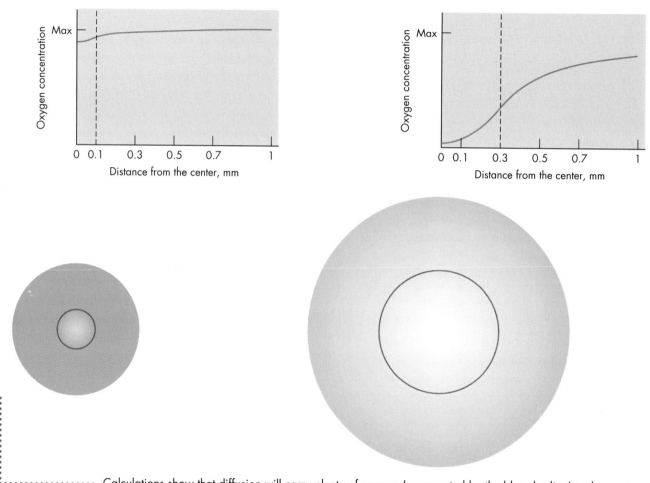

Calculations show that diffusion will carry plenty of oxygen (represented by the blue shading) to the center of a small, spherical organism measuring one-tenth of a millimeter in radius, but that oxygen will be depleted at the center of an organism only three times as large. The organism is assumed to consume 1 milliliter of oxygen (at standard temperature and pressure) per cubic centimeter of volume per hour. The oxygen concentration in the environment is assumed to be in equilibrium with the atmosphere, at 0.007 milliliters per cubic centimeter.

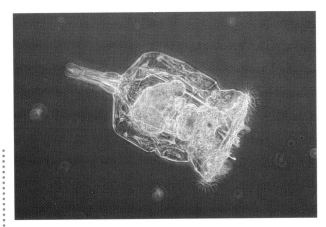

Rotifers, such as this *Brachionus,* are simple animals with hundreds of highly specialized cells, including nerves and muscles. They swim by means of a circle of small whiplike structures called cilia. The beating of the cilia sometimes creates the impression that the circle is rotating, and the effect has given rise to the rotifers' common name, "wheel animals." All rotifers are microscopic, less than a millimeter in size.

ciently high concentrations be maintained everywhere within the organism? The graphs on the opposite page show the results of calculations for spheres of radius 0.1 millimeter (abbreviated mm) and 0.3 mm. The calculation assumes that diffusion is the only mechanism of transport; if fluid inside or outside the organism is moving, the rate of transport would increase.

For an organism with a radius of 0.1 mm, the concentration is not reduced much below the maximum anywhere. Yet, for an organism only three times larger, oxygen is significantly depleted everywhere in the organism and nearly absent from its center. We conclude that organisms can rely on diffusion for transport only if they are smaller than some fraction of a millimeter. This conclusion fits with our observations of the creatures we find in nature. The tiny rotifers, frequently called "wheel animals" because

they generate currents for locomotion or feeding by means of a circle of waving cilia around their head, are composed of several hundred cells organized into well-defined organ systems, including muscles and a nervous system of several hundred neurons; but they lack a circulatory system, and their diameters are less than a millimeter. Nematodes, or "round worms," are similar to rotifers in size and organization but move by undulating their elongated cylindrical bodies; they also lack a circulatory system and they rarely exceed 0.1 mm in diameter. In contrast, arthropods (such as crustaceans and insects) have circulatory systems and commonly grow to much larger sizes.

Now we can return to the question of why the large animals familiar to us do not have rotary motion. A simple explanation is that a rotating appendage larger than a millimeter would need a circulatory system connected to the rest of the body in order to obtain nutrients and rid itself of waste products. Simply connecting blood vessels between the body and the appendage would not work, however, since the artery and vein would be twisted around each other as the appendage rotated. The twisting blood vessels would stretch until they broke or stopped the rotation. Limited rotation could occur without too much twisting, and this actually does take place in our hip and shoulder joints. But unlimited rotation would be impossible with a circulatory connection. Bacteria can have rotating appendages only because diffusion transports molecules without the help of the circulatory system required by large animals.

Varieties and Functions of Microbes

The size distinction between organisms with a circulatory system and those without one happens to fall near the visual threshold of the naked eye—about

0.1 mm. This size thus offers us a practical limit at which organisms could be considered microorganisms. In fact, I will include among microbes those simple animals that are sufficiently small that they lack a circulatory system. Thus, I discuss rotifers, nematodes, and flatworms as well as the fungi, protozoa, and bacteria traditionally considered to be microorganisms.

The smallest organisms are bacteria, about one millionth of a meter in diameter. If a hair from your head were enlarged to the size of a tree trunk, a typical bacterial cell would be the size of a cockroach. At this small scale, it's awkward to give sizes in meters or even millimeters, so the unit of measurement commonly used when talking about microbes is a micrometer, or millionth of a meter, abbreviated μm.

Besides being the smallest organisms, bacteria are also the simplest, surrounded by a membrane and cell wall, but lacking much structure inside the cell. Most obviously, the genetic material is spread throughout the cell, instead of being concentrated in a separate compartment, the nucleus. Bacteria are called prokaryotes (which means "before the nucleus") because they are thought to have evolved before organisms with a nucleus, which are called eukaryotes (meaning "true nucleus"). Fossils tell us that bacteria existed for two billion years before more complex cells appeared on earth. Bacteria are found anywhere that any other organism can be found and in many places where no other form of life can survive—such as in hot springs or salt-saturated ponds. Some are photosynthetic such as the cyanobacteria, which used to be commonly known as blue-green algae.

Molecular studies in the last two decades have led scientists to conclude that there are two fundamentally different types of bacteria—now called archaebacteria ("old bacteria") and eubacteria ("true bacteria"). Many archaebacteria are specialized to live near volcanic vents and in other hot environments lacking oxygen, which are thought to provide conditions similar to those on earth when life first evolved. The most familiar bacteria are eubacteria. These two types of bacteria differ in the molecular composition of their cell walls, membranes and protein synthesis machinery. In fact, in the composition of these basic elements, animals are more similar to plants than eubacteria are to archaebacteria.

All large organisms, as well as many microbes, are eukaryotic organisms. They store their genetic information in DNA molecules separated from the rest of

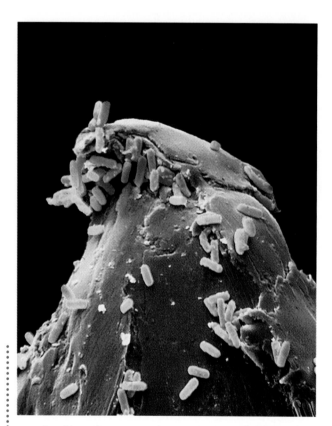

A colorized scanning electron micrograph (SEM) of bacteria (yellow) on the point of a syringe needle (blue), showing their minute size. The bacteria are about 5 micrometers in length.

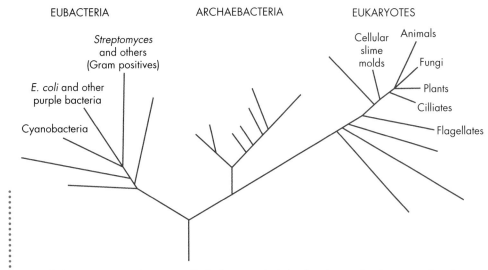

EUBACTERIA ARCHAEBACTERIA EUKARYOTES

Streptomyces and others (Gram positives)

E. coli and other purple bacteria

Cyanobacteria

Cellular slime molds

Animals

Fungi

Plants

Cilliates

Flagellates

DNA comparisons have rapidly led to new insights into how microbes are related to one another. In this summary of data, the length of the lines joining two types of organism is proportional to the number of changes that must have occurred in their DNA since they last shared a common ancestor. Surprisingly, plants and animals are more closely related by far than many of the microbes. Overall, the pattern suggests that there are three main types of organism existing on earth today—eubacteria (ordinary bacteria), archaebacteria, and eukaryotes (which probably formed when a symbiotic or parasitic association of bacteria eventually evolved into a complex cell with a nucleus, mitochondria, and sometimes chloroplasts).

the cell by membranes, forming a nucleus. Almost all eukaryotic cells also contain membrane-bound mitochondria that produce most of the energy for the cell. In addition, plant and algae cells contain membrane-bound chloroplasts that carry out photosynthesis. A variety of evidence suggests that all three of these structures originated from bacteria that invaded the cell billions of years ago. This organization has been very successful, and there are a wide variety of eukaryotes—algae, protozoa, fungi, plants, and animals. The individual cells of these organisms are generally larger than bacteria by 10 to 100 diameters, but they are still microscopic in size.

Algae are simple photosynthetic eukaryotes that lack the complex organization of plants. Most of them are not like the seaweed we find along the beach, but are microscopic single-celled organisms, some even capable of locomotion by the action of flagella. Algae include a variety of specialized forms that are not closely related to one another—some, like diatoms and dinoflagellates, armored with hard shells made of cellulose or minerals. As a group they are found wherever there is sunlight and moisture—even on moist soil and snow. They are especially abundant in the plankton of the oceans, and the algae there account for most of the photosynthesis on earth. Myriads of other single-celled eukaryotes—the protozoa—are not photosynthetic. They dart about by

A Note on Nomenclature

Traditionally, species have been defined as comprising all individuals that are capable of breeding with each other. The classification of microbes into species is often difficult because many reproduce asexually (by division without the union of parental cells). Consequently, individuals give rise to clones of nearly identical offspring, and the usual definition of a species as an interbreeding population is difficult to apply. Nevertheless, the standard naming system is so well established that microbiologists force their classification schemes to fit the species model established for large organisms. In practice, taxonomists have used the size and shape of the organism, the characteristics of colonies formed on agar, and metabolic behavior to identify species, but analysis of DNA sequences is rapidly becoming the best method.

In the basic scheme for classifying organisms, the genus is the next most inclusive grouping above the species level. A typical genus contains about 10 species, although a genus may include anywhere from one to dozens of species. Particular species of organisms are usually identified by giving the name of the genus to which the species belongs followed by the name of the species. The genus name is normally capitalized, while the species name is not. Different species of microorganism within the same genus often cannot be distinguished unless examined by special techniques, whereas members of different genera are relatively easy to distinguish and often correspond to the distinctions between organisms made by nonscientists. For microbes, genus names are often used as common names. When separate populations of microbes have a few distinctive characteristics but are still considered to belong to the same species, they are often referred to as different strains.

As an example of the classification system, consider some familiar organisms of the Order of carnivores belonging to the Class of mammals:

Genus	Species	Common name
Canis	*familiaris*	domestic dog
Canis	*lupus*	gray wolf
Canis	*latrans*	coyote
Felix	*domesticus* (or *catus*)	house cat
Felix	*leo*	lion
Felix	*tigris*	tiger
Ursus	*americanus*	American black bear
Ursus	*ferox* (or *horribilis*)	grizzly bear
Ursus	*maritimus*	polar bear

The meaning of traditional terms like bacteria, algae, fungi, and protozoa has become especially confused in recent years, as the results of DNA sequencing greatly improve our understanding of how organisms are related to one another. The results clearly indicate that the organisms traditionally included within these groups are not defined by sharing a recent common ancestor. Nonetheless, it is convenient to use these terms to indicate organisms sharing certain characteristics without implying that they form a coherent group related by evolutionary descent.

means of cilia or flagella, or crawl like amoebae, in search of something to eat.

An advantage of being small, in addition to not needing a circulatory system, is that special structures like gills to expand surface area are not required. The surface-to-volume ratio decreases as any object increases in size. Thus, a larger organism has less surface through which materials can pass for a given volume of the organism, and yet a larger organism needs to import just as many nutrient molecules per unit volume. Because of this limitation, large organisms must have specialized structures that increase the surface area to increase the uptake of nutrients—lungs, gills, and leaves are examples. These structures can increase surface area enormously; the surface area of the human lung is about 40 times the external surface area of the body. But, because of their small size, most microbes have no need for specialized structures such as these.

Fungi are exceptional because, although simple in structure, some of them grow to become quite large. Fungi are fundamentally microorganisms because they

This "fairy ring" of toadstools in a lawn in Tulsa, Oklahoma, demonstrates that fungi can grow to quite large sizes. The fungus started growing in the center underground, forming a network of fine threads—called hyphae—that digest organic matter and absorb the released nutrients. At a certain time of year, if conditions are favorable and the fungus has prospered, it reproduces by forming spores that are released from the "toadstools" into the air for dispersal. Most of the toadstools form at the periphery of the hyphal mass where the fungus has access to fresh organic matter; thus the extent of the ring indicates how far the fungus has spread.

usually consist of cylindrical filaments only about 10 μm (0.01 mm) in diameter. In many fungi, these filaments may grow into a large network called a mycelium, which in some cases can extend several meters and weigh pounds. Such huge mycelia are possible because this type of organization—a network of small filaments—provides the mycelium with tremendous surface area. A cylinder of fixed diameter can grow to any length without decreasing its surface-to-volume ratio. This large surface area allows fungi to specialize in external digestion: the filaments release digestive enzymes that break apart organic matter, then absorb the nutrients freed by digestion. In addition, fungi are able to guide the growth of their filaments toward sources of nutrients so that the nutrients can be obtained by diffusion over very small distances. In a sense, fungi use directed growth to make up for their lack of locomotion.

It is no exaggeration to say that without microbes the familiar forms of life could not exist. This is not simply because larger organisms and microbes evolved from a common ancestor, which was itself a microbe. Even today, all of life is dependent upon the "services" performed by these invisible organisms. One of the most important of these services is to provide the energy needed to carry out the metabolic activities necessary to all living organisms.

All organisms take in nutrient molecules of various kinds, rearrange the atoms of these nutrients to form thousands of new kinds of molecules, and eventually excrete most of the atoms as waste products. The reorganization of atoms from one molecular arrangement to another sometimes requires energy to be added and sometimes causes energy to be released; moreover, the process can be reversed. For example, hydrogen ($2H_2$) and oxygen (O_2) can combine to form water ($2H_2O$), giving off energy (as burning hydrogen gives off a flame), and water can be split with

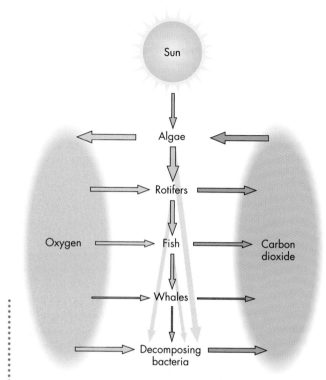

Many atoms are cycled through a series of organisms, whether the atoms are ingested in food or breathed in like oxygen, whose passage through such a series is illustrated by this depiction of the oxygen cycle. Oxygen atoms in the atmosphere (O_2) are taken up by animals and used to "burn" food, producing energy and the waste products carbon dioxide (CO_2) and water (H_2O). Photosynthetic organisms, like algae and plants reverse the process, using energy from the sun to convert carbon dioxide to food, releasing oxygen as waste. Similar food chains occur for other nutrients.

the addition of energy to form hydrogen and oxygen (as takes place in electrolysis). The energy gained or lost is exactly the same amount for transformations in opposite directions, and we can think of energy as being contained within the molecules as potential energy.

A fundamental requirement for life is that the inputs (nutrient molecules and, in the case of plants and algae, light) must contain more concentrated energy than the waste molecules, and the energy difference is used by the organism to organize the atoms that remain in the body. After being put to work organizing the organism, the energy is degraded into heat and lost to the environment.

Most organisms specialize in using one of two kinds of energy input. Light is the source of energy exploited by photosynthetic organisms, called photoautotrophs, which can make the basic organic molecules found in all organisms out of simple carbon dioxide and water. Organisms like ourselves, called heterotrophs, must obtain their energy by eating other organisms. Because there are no true plants (only algae) in the oceans, which cover most of the earth's surface, bacteria and algae convert more carbon dioxide from the atmosphere to energy-supplying organic molecules than all the true plants combined. The "primary producers" become the foundation of food chains leading to the large consuming organisms we are most familiar with.

For centuries, it was thought that photoautotrophs and heterotrophs were the only types of organism. It has now become clear, however, that

This scanning electron micrograph shows the surface of a microbial mat, a conglomeration of microbes that form their own ecological community in a layer only a few millimeters thick. These bacteria are Thiothrix chemoautotrophs that obtain energy from mineral solutions emanating from hot springs—or from hydrothermal vents at the bottom of the ocean.

there are bacteria living in special habitats that obtain energy from chemical reactions. Examples of such "chemoautotrophs" are the bacteria living in animal guts, under swamps, and in other habitats lacking oxygen. Some of these bacteria combine hydrogen with carbon dioxide to produce methane as a waste product and are a cause of "marsh gas" and flatulence.

In recent times, it has become appreciated that microbes are essential to the cycling of nutrients in ecosystems. For example, all organisms require organic molecules containing nitrogen atoms, but only certain bacteria can convert nitrogen in the atmosphere to forms that are accessible to other organisms. Without the "nitrogen-fixing" activities of these bacteria, nitrogen would eventually become locked up in the atmosphere and life as we know it would cease.

Certain microbes are essential to our own health. Shortly after birth, we were all infected by a variety of bacteria—some colonize our skin and hair but most live in our lower intestine. Indeed, our bodies typically contain more bacterial cells (10^{14}) than human cells (10^{12}). Using special laboratory techniques, scientists have isolated "germ-free" animals without any microbes, but these animals are not healthy. They must be supplied with high amounts of vitamin K and certain B vitamins that are normally synthesized by intestinal bacteria, and they are unusually susceptible to infection by pathogenic bacteria. Apparently the bacteria normally living in the body help prevent the pathogenic types from prospering. Thus, most of the bacteria we carry around with us are beneficial.

Hunting and Farming: Two Life Styles in the Micro World

Considering the simplicity of microbes—the lack of muscles or a nervous system in any except the micro-

animals (rotifers, nematodes, and flatworms)—it may seem surprising that they exhibit any behaviors at all. Nevertheless it was their behaviors that first revealed to the earliest observers that these minute objects were alive. As we shall see, many microbes have developed surprisingly sophisticated behaviors, highly tuned to solving the problems they face and adapted to the constraints of physics at their small size scale.

One way of looking at behavior is to recognize that behavioral actions change an organism's environment, either by moving the organism to a new environment or by modifying the old one. That movement is a form of behavior important to microbes is immediately obvious to anyone who has used a microscope to observe them darting and crawling around. Movement is often an efficient mechanism for obtaining nutrients, light, or a mate. Although microbes cannot move far under their own power, they don't have to: the chemical environment often changes sig-

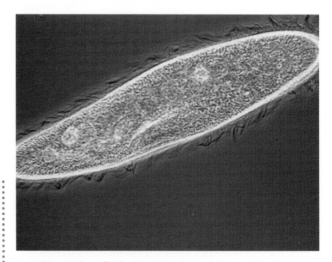

This single-celled eukaryote *Paramecium* swims by means of the cilia that cover its surface, some of which can be seen at the edge of the cell.

nificantly over distances of only a millimeter. Another kind of behavioral activity—secretion—is less obvious but no less important. Fungi secrete digestive enzymes (proteins that catalyze the chemical decomposition of organic molecules) to increase the supply of nutrients and, in some cases, secrete toxic chemicals to reduce competition from other organisms. In fact, many of the drugs we use to control bacterial infections are derived from bacteria and fungi, which synthesize them in order to eliminate their competitors.

Through movement a microbe, like a hunter, places itself in a more favorable environment; through the secretion of chemicals a microbe, like a farmer, modifies the environment that it is already in. Both types of behavior have costs: a microbe relying on locomotion expends energy to overcome friction; one using secretion requires energy and materials to synthesize the molecules it secretes. To a great extent microbes specialize in either locomotion or environmental modification but not both. Fungi cannot change location but do change their environment by secreting digestive enzymes that decompose organic matter. In contrast, some bacteria, many protozoa and algae, and most animals have mechanisms for locomotion. This specialization makes sense because it

would be wasteful for these microbes to invest in modifying the present environment only to abandon it and move to another.

We can see an example of the importance of locomotion by comparing two kinds of algae: diatoms and dinoflagellates are both important photosynthetic eukaryotes in the plankton of the sea, and both are in the same size range (2 to 2000 μm) and are surrounded by hard armor—crystallized silica in the case of diatoms and plates of cellulose in the case of dinoflagellates. The main distinction is that dinoflagellates are motile. But with that sole advantage dinoflagellates can outcompete diatoms in calm surface waters where there is little mixing to carry diatoms between deep sources of nutrients and light at the surface. Closer to home, locomotor abilities seem to be essential for many pathogenic bacteria to invade human cells and cause disease.

What other mechanisms have microbes invented for locomotion? How do they guide locomotion in the appropriate direction? How is secretion used in digestion, trading for food, or to suppress competitors? Do microbes communicate? Are they capable of learning? In this book, we will look at the various behaviors that allow microbes to prosper.

Some single-cell eukaryotes, like this *Paramecium,* are covered with thousands of short cilia that wave back and forth, pushing on the surrounding water and propelling the organism along.

Locomotion Without Legs

Although many microbes are motile, they move in ways that may seem strange to us in the large-scale world. None of them use the legs and wings of familiar animals, for in their peculiar world other mechanisms are more effective, including wavy "oars" and flowing cytoplasm. Much of the reason has to do with the behavior at small scales of the fluids through which microbes move. To a microbe moving in the ocean, the properties of the surrounding water are very different from those experienced by a much larger fish swimming nearby.

Viscosity Dominates

For a solid object like a microbe to move through a stationary fluid, the fluid must be made to flow around the object in order to make space ahead of it and to fill the space behind. A fluid doesn't always move out of the way easily, as you soon notice if you try wading through water for any length of time. In fact, an object meets two kinds of resistance to this flow. When a moving object approaches a particular part of the fluid, that part must be accelerated—first to flow out of the path ahead of the object, then to flow back into the path behind the object, then to stop. Acceleration is resisted by the inertia or mass of the fluid. You can literally feel this resistance when you stick your hand out the window of a speeding car or paddle a boat. In addition, the flow forces parts of the fluid to slide past one another. The sliding of part of a fluid past another part is resisted by friction, generating a fluid property called viscosity.

The inertia of a fluid, the property that causes it to resist acceleration, is proportional to its density. This density is usually symbolized by the Greek letter ρ (rho) and defined as the mass per unit volume. Water has a density very close to 1 g/cm^3, while air is a

thousand times less dense. This difference in density is the reason that raindrops fall and air bubbles rise, and it also explains "belly flops" into water. A human falls through air rapidly because the low-density air easily moves out of the way, but the fall is abruptly slowed once the diver reaches the water surface, because to move the denser water out of the way requires much more force. This greater resistance of water can be put to use in paddling a boat, but try to move a wagon by paddling air—high-speed propellers are required.

Viscosity is a measure of how strongly one part of a fluid resists motion relative to another. When one part of a fluid is in motion, it tends to drag adjacent parts along with it. But if a part is being dragged in different directions by different neighbors, something must give, and one part slips with respect to adjacent parts. The slipping converts mechanical energy to heat as happens in other forms of friction. Viscosity is thus a measure of the resistance to slipping. It is symbolized by Greek letter η (eta) and is measured by determining the force necessary to move a solid object through the fluid or to move the fluid through a tube. For example, the force necessary to move a sphere through a fluid at low speeds is equal to the product of:

- 6π,
- the radius of the sphere r,
- the velocity v of the sphere relative to the medium, and
- the viscosity η of the medium.

This relationship is known as Stoke's law, which states: force = $6\pi r v \eta$.

Air and water differ in their density and viscosity, and thus offer different resistance to a moving object. They also differ in compressibility, a measure of how much the volume of a fluid changes with a change in pressure. Gases are highly compressible, while liquids

Properties of Air and Water

Medium	Density, g/cm^3	Compressibility, fractional volume change/bar[a]	Viscosity, g/s cm[b]
Air, standard	0.0012	1	0.000179
Pure water, 0 °C	0.9998	5.1×10^{-5}	0.01787
Pure water, 20 °C	0.9982	4.6×10^{-5}	0.01002

[a]A "bar" is a unit of pressure approximately equal to typical atmospheric pressure.
[b]Viscosity has units of mass per length per time.

are barely so. Air behaves much like an ideal gas, which compresses to half its original volume when pressure is doubled. In contrast, water subjected to the same pressure change is compressed by only about 50 parts per million of its original volume. In fact, it is usually assumed that water and water-filled organisms are incompressible, for the compressibility is so small that it usually makes no difference.

When a fluid flows sufficiently slowly past a solid, its behavior is simple and predictable. Each particle of the fluid moves nearly parallel to its neighbors. The fluid adjacent to the solid "sticks" to it, but adjacent layers slip; fluid close to the solid's surface flows nearly parallel to the surface. This smooth pattern of flow is called "laminar" flow. For a sphere moving through a fluid, the fluid moves nearly parallel to the surface of the sphere as the fluid moves out of its path during approach of the sphere and back toward its path after it has passed.

As speed increases, however, the pattern of flow changes more and more sharply. The fluid flowing around the moving sphere is more abruptly accelerated—first to move out of the path, then to move back toward the path, and finally to stop in the path behind the sphere. At some speed, the inertia of

the fluid is too great for the fluid to change direction quickly enough, and the pattern of flow breaks up to form waves and vortices. This change in the pattern of flow is the onset of turbulence. In fully developed turbulence, the patterns of flow are fundamentally unpredictable, but it is possible to predict when turbulence will develop by knowing a quantity called the Reynolds number.

The transition to turbulence is best understood as accompanying changes in the relative importance of viscosity and inertia. The fluid has plenty of time to move out of the way of a slowly moving object: its acceleration is low, so inertia is not important, and viscosity minimizes the slipping of one layer of fluid past its neighbors. We say that viscosity dominates inertia in such laminar flow. As the moving object's speed increases, so does the fluid's rate of acceleration; inertia, which resists acceleration, becomes more important. Eventually, the laminar pattern of flow requires such rapid changes of speed and direction that the fluid cannot follow it, and turbulence breaks out. It is the relative strength of the inertial forces to the viscous forces that is captured by the Reynolds number.

One of the most important concepts in studying flow, Reynolds number is the ratio of forces due to

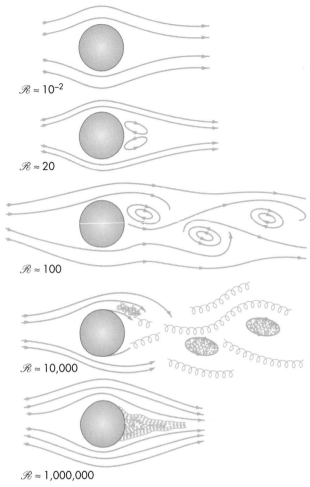

$\mathcal{R} \approx 10^{-2}$

$\mathcal{R} \approx 20$

$\mathcal{R} \approx 100$

$\mathcal{R} \approx 10,000$

$\mathcal{R} \approx 1,000,000$

Patterns of fluid flow around a cylinder are different at different Reynolds numbers ($\mathcal{R}$). At low speeds, where $\mathcal{R}$ is less than one, the flow is smooth and laminar. At higher speeds, vortices form and the flow breaks up into chaotic patterns that define turbulence.

inertia to forces due to viscosity. The forces due to inertia are equal to the product of:

- the density of the fluid (which is a measure of its inertia),
- the length of some relevant structure such as the

diameter of a moving sphere (which measures the distance the fluid must move to get out of the way), and

- the speed of the fluid with respect to the structure (which measures how quickly the fluid must get out of the way).

To obtain the Reynolds number, this product is divided by the viscosity of the fluid (which measures the frictional resistance to sliding). All the units cancel out, producing a dimensionless number. Its importance derives from the fact that, for a specific surface shape, at a particular value of the Reynolds number, the pattern of flow past the surface is the same independent of the individual values of density, length, speed, and viscosity. It is this similarity of flow pattern at different scales that allows engineers to accurately test designs of ships and airplanes using scale models.

Stoke's law giving the force necessary to propel a sphere through a fluid is only valid for laminar flow and always applies when the Reynolds number is well below 1. For higher values, however, turbulence causes the resistance to increase by more than Stoke's law predicts, and no simple theory can predict what that resistance will be. The observed increase in resistance over what Stoke's law predicts is about 10 percent at a Reynolds number of 1, 60 percent at 10, and 500 percent at 100.

Depending on the geometry of the flow pattern, the transition to turbulence occurs under conditions in which the Reynolds number is in the range 1 to 10,000. At higher values, inertia dominates, and turbulence is common; at lower values, viscosity dominates, and flow is smooth and predictable.

As you can see in the table on the facing page, the Reynolds number varies enormously for swimming organisms—by 13 orders of magnitude. This huge range can be explained by the fact that organisms vary over a

Swimming at Different Sizes

Organism	Size, cm	Speed, cm/s	Reynolds number
Whale	10^3	10^3	10^8
Human	10^2	10^2	10^6
Fish	10	10^2	10^5
Copepod	1	10	10^3
Rotifer	0.03	0.1	0.1
Paramecium	0.02	0.1	0.1
Bacterium	10^{-4}	10^{-3}	10^{-5}

millionfold in size, speed is roughly proportional to size, and the Reynolds number is proportional to the product of size and speed. It is also clear from the table that microbes live in a world with very low Reynolds numbers. Consequently, we may be certain that they do not experience turbulent flows. In contrast, almost all other animals—including our-selves—contend with high Reynolds numbers and turbulence.

The extreme differences between the world of the microbe and our familiar world can be illustrated by asking the question: If a swimming bacterium suddenly stopped propelling itself, how far would it coast? A simple calculation predicts that it would coast much less than the width of a single atom—a millionth of its length. This reflects the dominance of viscosity over inertia.

Another approach is to ask what the viscosity would have to be for an organism on our size scale to experience the same Reynolds number as a bacterium. The answer is about 10^{11} times the viscosity of water. If air had this viscosity, a human body would fall through it at a speed of only one millimeter per hour! Thus, we are safe in saying that inertia has no relevance to microbes; their world is dominated by viscosity.

A surprising feature of locomotion at low Reynolds numbers is that streamlining is often coun-terproductive. At high Reynolds numbers, the drag (resistance to movement through a fluid) of an object

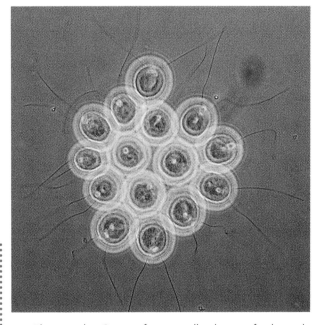

The microbe *Gonium* forms small colonies of spherical cells, each having a pair of flagella. The cells of a colony (usually numbering 4, 8, 16, or 32 cells) are arrayed side by side in a single nearly flat layer; all the flagella point away from the same side of this layer. With the flagella leading, the colony swims with its broad side facing forward, so that the broadest possible area is oriented perpendicular to the movement. Although this orientation would be very inefficient for large objects, in the micro world the shape and orientation make very little difference.

can be reduced by adding surface area to make the overall shape more streamlined. A disk, for example, can be made more streamlined by expanding it to form a sphere of the same diameter. In contrast, at low Reynolds numbers, drag has little dependence on shape. For example, at Reynolds numbers over a thousand, a hemispherical shell has about four times less drag when moving toward its convex side as toward its concave side, while at low Reynolds numbers there would be no difference. The object's shape does not matter so much at low Reynolds numbers, because under these conditions an object moving through a fluid drags much more of the neighboring fluid with it than at high Reynolds numbers.

Another important feature of life at low Reynolds numbers is that flow is reversible and mixing is difficult; if you stir one solution into another, you can un-stir them by reversing the stirring motions. If a drop of dye is placed in a layer of glycerol between two concentric cylinders, and one cylinder is rotated, the dye will be stretched out into a thin sheet that can wrap around the inner cylinder several times. If you look through the layer of glycerol, it looks as if the dye has mixed into the original glycerol. If the rotation is reversed, however, the dye will return to one spot. This amazing demonstration makes it superficially appear that time has been reversed, since we all know that mixing progresses toward uniformity.

Because of this reversibility, a simple back-and-forth motion of a rigid structure cannot produce loco-motion—it simply moves an organism back and forth over the same short path. A scallop, for example, swims by opening and closing its two rigid shells. It makes this motion propulsive by imposing an asymme-try in speed, opening the shells slowly and closing them rapidly. The rapid closing of the shells produces a jet of water whose inertia propels the scallop back-ward. But a micro scallop living in a low Reynolds number world could not swim this way because the in-ertia would be insignificant.

The Harvard physicist Edward Purcell calls this conclusion the scallop theorem. Purcell describes swimming as locomotion through water produced by repeating a pattern of changes in body shape over and over again. Some cyclic patterns, like the scallop's, are reciprocal in that one half of the cycle includes the same shapes as the other half but in reverse order (A→B→C→B→A→); a snapshot of the or-ganism would show a shape that could be in either of the cycle's two halves. Other cyclic patterns are not

The shapes of a swimming organism are represented by the values of one or two parameters (such as the angle of a joint), which vary to produce shapes A, B, C, D. According to Purcell, reciprocal patterns such as those in the top two lines, in which the same path of shape changes is traversed in opposite directions (A→B→C→B→A→. . .), will not produce any net change in position at low Reynolds number, at which flow is reversible. The organism will be at exactly the same location each time it returns to a particular shape (such as A). However, an organism changing shape in a nonreciprocal cyclic pattern (A→B→C→D→A→. . .), such as in the bottom row, will be at a different location each time it gets to a particular shape. Such an organism will be able to swim forward.

reciprocal because each shape occurs in only one half of the cycle; a swimmer returns to the shape its body held at the start of a cycle by going through a different series of shapes from those it went through when it departed from the starting shape (A→B→C→D→A→). Only nonreciprocal motions succeed in moving an organism at low Reynolds numbers. Any nonreciprocal change in shape will produce net displacement and rotation, which repeated over many cycles will drive the organism along a helical path. Some patterns of change propel an organism much more efficiently than other patterns, but it is difficult to determine which patterns would be most efficient. Physical theory provides little help as the hydrodynamic flows are too complex for us to draw simple general conclusions. What is observed is that swimming microbes of different types employ very different kinds of shape changes.

Propellers

The simplest mechanism for locomotion at low Reynolds numbers is also the common mechanism used to propel boats with engines: a propeller or screw that rotates continuously in one direction. Many bacteria make use of a rotating screw, called a flagellum, composed in the simplest cases of a protein crystal forming a thin filament in the shape of a helix. In the best-studied bacteria, the flagellum is about 10 μm long and only 0.02 μm in diameter, far too thin to be resolved by the light microscope, although its presence can be visualized by light scattering. In one case, it has been estimated that the flagella contained 8 percent of all the protein in the cell—a sizable investment that suggests how important locomotion can be to bacteria.

The filament is rotated by a biological motor at the site of attachment to the cell. This motor can turn

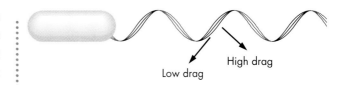

Low drag   High drag

Bacterial flagella, drawn here to scale, cluster at one end of the cell in some species, although in others they may cluster at both ends or (as in *E. coli*) form on the cell surface at random locations. Each filament is a protein "crystal" with a diameter of only 0.02 μm twisted into a helix with a diameter of 0.5 μm. When the flagella rotate in the appropriate direction, the forces of viscosity cause them to form a bundle, which acts like the crew of a ship to push the cell forward. When they rotate in the opposite direction, the bundle comes apart, and each flagellum pushes the cell in a different direction, sending it into a chaotic tumble.

faster than 100 times a second, or 6000 rpm—about the top speed of a car engine. As the only example of rotary motion in biology, it has received a great deal of attention from researchers.

Rotating the screw generates a propulsive force because the viscous drag on a filament is lower for displacement along its length than sideways. Investigators have measured the drag on straight wires placed in viscous solutions to duplicate the conditions of low Reynolds numbers. The drag on a wire moving parallel to its length is about a factor of 2 less than the drag on a wire moving perpendicular to its length. Now in a helical filament rotating around its axis, each segment of filament is moving around the axis but oriented at an angle to it. As it pushes against the fluid, the segment tends to slide along its length because of the lower drag in this direction. The net effect of the sliding of all the segments of the helix is to push the helix in one direction along its axis. However, since the difference in drag is only about a factor of 2, there is a great deal of slipping, and the helix does not advance as far as it would if it were pushing against solid structures.

In one study, bacteria advanced only 0.18 μm for every rotation of the filament, or only about 7 percent of the 2.5-μm advance that they could have achieved had there been no slipping. Purcell estimates that only about 1 percent of the energy expended in propulsion actually contributes to forward motion. Still, bacteria can swim faster than 10 body lengths per second at an energy cost amounting to only a small fraction of the total energy generated by their metabolism.

Another factor reducing efficiency is that, because of Newton's law of action and reaction, the force rotating the filament causes the cell to rotate in the opposite direction. In one carefully studied case, the cell rotated about 7 times a second, while the filament rotated 100 times a second in the opposite direction. Thus, the filament rotated about 14 times for every rotation of the cell. If the cell is straight, its rotation wastes energy, which may explain why some bacterial cells have a helical shape.

Observed through the electron microscope, the structure rotating the filament—its motor—is revealed as a complex structure inserted in the cell wall and membrane. The motor's mechanism has proved difficult to study. An electron microscope cannot observe the motor in action because this instrument can only reveal structures remaining in a vacuum, devoid of water, yet, at a size of only 0.03 μm, the motor is much smaller than the limit of resolution of the light microscope. On the other hand, the structure is very large for study by the techniques of isolation and purification used to study single molecules. Nonetheless, a great deal has been learned about the motor's 20 kinds of protein molecule, as well as the 30 genes required for its assembly and function, using electron microscopy, protein purification, and genetic muta-

tions, mostly applied to the intestinal bacteria *Escherichia coli* and *Salmonella typhimurium*.

The filament is attached to the motor by a structure called the "hook," which functions to change the orientation between the axis of the filament and the motor. Researchers speculate that the hook may also provide flexibility. Like the filament, the hook is composed predominately of a single type of protein molecule—about 120 copies of that molecule are present. The hook attaches to a rod that is surrounded by two rings in some bacteria and four rings in others. The rings are thought to interact with specific layers of the cell covering, including the cell wall and the inner

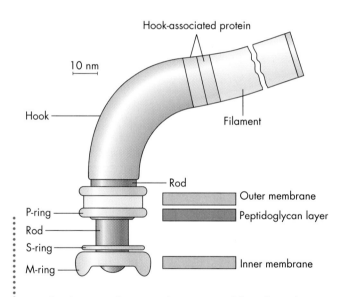

A schematic diagram of a motor and flagellum of *Salmonella typhimurium* bacteria. (The full length of the filament is actually hundreds of times its diameter.) Each of the structures identified is formed from many molecules of a particular kind of protein. The hook rotates and is thought to act as a flexible joint. Researchers assume that the rings are anchored to the cell wall, and the rotational force is generated between them and the rod, which is connected to the hook.

membrane. Yet even when four rings are present, only two are essential for rotation. Unfortunately, no one has yet figured out a way to determine which rings rotate and which are stationary, or to answer the basic question of where the rotational force is generated, so the functioning of this motor is still mysterious.

The performance of the bacterial motor is also difficult to study, but progress has been made using several innovations. In particular, the physicist Howard Berg has focused his efforts on understanding the bacterial motor. Because a single filament is impossible to see while it is moving, Berg studies its rotation by attaching the filament or hook to a large solid, by this trick making the cell rotate instead of the filament—the tail wagging the dog, so to speak. Because the cell body has more viscous drag than the filament, the cell rotates slower (up to 15 times a second) than the filament in a free cell (up to 100 times a second). At the slow speed of a tethered cell, a motor is operating close to its slowest speed without stalling and is very efficient. At least 50 percent of the electrochemical energy it uses is converted to the mechanical energy of rotation. However, at the high speeds at which flagella rotate in free-swimming cells, friction in the motor reduces the motor's efficiency to roughly 5 percent.

In the most elegant experiments, the experimenters have applied forces to make the cell rotate at controlled speeds. To create the forces, Berg and his collaborators use strong gradients of intense light from a laser to make a so-called optical trap, or they apply rapidly rotating, strong electric-field gradients. By making use of these "handles" on the cell, the scientists were able to estimate the forces generated by the motor at different speeds and directions of rotation. They can even supply sufficient force to make the motor slip and break. Their initial data have aided scientists in testing various proposed models for how

the motor works, but no model clearly stands out as superior.

For additional evidence of how the motor functions, investigators have turned to the techniques of modern genetics. One strategy has been to experiment on cells containing a mutation that inactivates a gene encoding a certain membrane protein. The experimenters introduced artificial chromosomes containing a functional copy of the gene into cells with the defective gene. Although the gene was not operating when first inserted, it was later turned on by adding a chemical. The investigators then measured the rotation of the cells as the gene began to function. The motor gained strength in a series of steps, as if the force-generating machinery were built of a series of equivalent units, each of which was assembled and put into operation in turn. A motor can apparently have a maximum of eight such force-generating units.

We may not know precisely what the rotary motor is made of, or how it works, but we do know how it is powered. Numerous experiments, on a variety of bacteria, have shown that the power source is one of the most common in biology—the movement of ions across a membrane. A difference in the concentration of any chemical represents a store of potential energy, as in a battery. Cells often use differences in the concentration of certain ions on opposite sides of a membrane as a store of energy and a means of transporting energy from one place or process to another—just as we use batteries. As ions move down a concentration gradient across the membrane, they can be made to do work. Eukaryotic cells, for example, use proton gradients in making ATP, a molecule that serves as the cell's energy store. In most cases studied, the ion whose movement powers a bacterium's flagellum is the hydrogen ion—a hydrogen atom from which the single electron has been stripped away, leaving a proton. Roughly 1000 protons cross the cell membrane to power each rotation of the motor—in only 0.01 second. As we shall see in Chapter 4, an important feature of the motor is that it is reversible—it can turn in either direction.

The single filament that propels some bacteria is replaced in many other bacteria by a variety of more complex arrangements. Cells often have several filaments, in some cases inserted at one or both ends of the cell (polar flagella) and in other cases inserted at seemingly random locations over the cell's surface. When the randomly placed flagella rotate in the appropriate direction, however, hydrodynamic interactions cause the flagella to bundle together and act as a single screw. Cells with these randomly dispersed flagella swim more effectively in viscous solutions such as those within our guts than cells with polar flagella, while polar flagella are more effective in the less viscous water of ponds and oceans. Some bacteria can switch from one type to the other as conditions warrant.

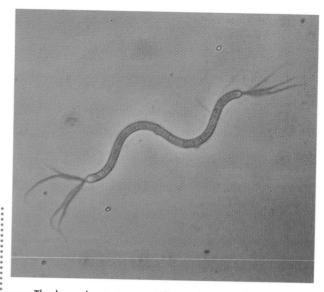

The large bacterium *Spirillum volutans* is frequently found living at low oxygen concentrations in stagnant fresh water. The cells of these bacteria are about 1.5 μm in thickness but grow in a helical shape up to 60 μm long. This specimen has been stained to make the multiple flagella at both ends visible in the light microscope. The helical cell shape helps the bacteria swim, as the cell rotates opposite to the flagella.

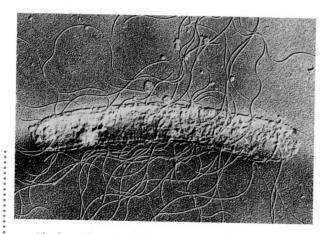

The large bacterium *Proteus mirebilis*, 20 μm long, has dozens of flagella attached all over its surface, as seen in this colorized electron micrograph. When swimming, the flagella form a single bundle, which acts like a ship's propeller.

It has long been observed that some bacteria are long, thin cells with a helical shape; some of these cells are over 100 times longer than they are wide. Such a helical shape would advance through the surrounding medium if rotated; but, in order to rotate, the cell must have something to push against. The bacteria previously described rotate their helical flagella by pushing them against the cell body, causing it to rotate in the opposite direction. Cells of freshwater-living *Spirillum* and ulcer-causing *Helicobacter* bacteria are rigid helices from which project rotating flagella. The action of the flagella causes these cells to rotate in the opposite direction and screw themselves through the medium. This ability may assist pathogenic species in invading the body.

More mysterious are the spirochetes, some of which cause syphilis or other diseases. Although their flexible cells are helical in shape, their flagella remain inside the sheath formed by the bacteria's outer membrane (in the so-called periplasmic space). Nonetheless, they swim very well and, unlike most other bacteria, move through viscous gels without slipping or becoming stuck in the gel.

Observation and experiment haven't told us much about how spirochetes propel themselves—only that these organisms have structures similar to bacterial motors, so we assume their flagella rotate as in other bacteria. A theoretical analysis of the problem has led to a proposed solution. It has been suggested that a helical cell could propel itself if its surface could be made to rotate around the axis running down the middle of the cell body. The rotating flagella might cause the outer membrane sheath to rotate with respect to the inner rigid helix. This rotation drags on the water surrounding the cell and causes the helical shape to rotate in the opposite direction. Rotation of the helix could drive the spirochetes through the medium. This remains a very attractive hypothesis, but it still awaits detailed supporting evidence.

An elegant solution to the problem of providing a counterrotating force would be to couple a right-handed helix with a left-handed helix and rotate one helix with respect to the other. In this situation, both parts of the cell would act in concert to drive the cell through the medium. It has recently been demonstrated that the spirochete *Leptospira illini,* which has a flagellum at each end, probably makes use of such a strategy indirectly. When the flagellum at either end rotates in one direction, it apparently causes that end of the cell to change from a right-handed helix (diameter = 0.2 μm; pitch = 0.7 μm) to a left-handed helix

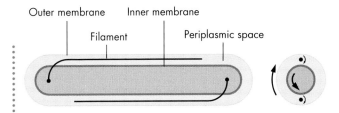

Outer membrane   Inner membrane

Filament   Periplasmic space

A schematic diagram of a spirochete bacterium, which is normally longer and bent into a helical shape. In these bacteria, the filaments of the flagella are contained in the periplasmic space and do not penetrate the outer membrane into the external environment. These organisms are self-propelled screws, caused to rotate through the action of the flagella.

(diameter = 0.6 μm; pitch = 2.7 μm) and gyrate. This end of the cell develops a pseudorotary motion that causes the rest of the cell to rotate in the opposite direction.

Bacteria with helical bodies seem to be particularly good at moving through gels and solutions of highly elongated molecules. These environments provide a relatively rigid structure against which the cells can push, and the cells propel themselves without slipping. Parasitic species commonly encounter such structured environments in the bodies of their hosts, and other species encounter them in decaying organisms. In contrast, the basic laws of hydrodynamics indicate that slipping is necessary to sustain movement in any simple solution, because force can only be generated by moving the fluid, and organisms traveling in open water must live with the inefficiencies of swimming.

Oars

Another simple mechanism for propulsion at low Reynolds numbers is the movement of an oar. It may seem surprising that an oar could be used as a means of locomotion in the micro world, since a microbe surrounded by water can't remove its oar from the water during the return stroke. Yet oars are quite common among microbes, and are relied upon by many algae, protozoa, and other single-celled eukaryotes. But there is a key difference between the stiff boards plied by human rowers and the living appendages used as oars by microbes: the microscopic oar is flexible and bends. More specifically, it bends in one direction during part of the cycle of movement and bends the other way while retracing its position. Consequently, the shapes are not exactly reciprocal, and locomotion is produced.

The oars of eukaryotic cells are thin protrusions, about 0.2 μm in diameter, called cilia or flagella. Despite the identical names, the flagella of eukaryotes should not be confused with those of bacteria, which function in a very different manner. Individual cilia and flagella closely resemble each other in molecular organization but are usually distinguishable, and cells never have both. Observing a cell under the microscope, one usually sees either one or two long flagella (100 to 200 μm) bending and waving to move fluids toward or away from the cell, or one sees arrays composed of hundreds or thousands of short cilia (5 to 20 μm), waving almost in synchrony to move fluids across the cell surface. These two structures are also important in large multicellular organisms like ourselves: flagella give sperm their motility, and cilia keep our airways clear of fluid and debris.

In the simplest cases, regions of bending move along the length of the cilia or flagella to produce a

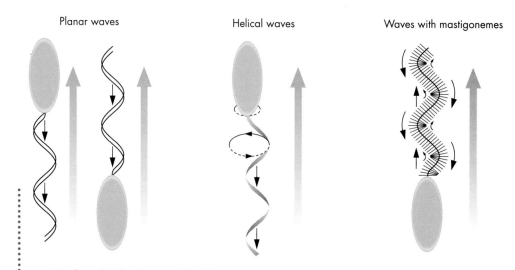

The flagella of eukaryotes bend in a variety of ways, unlike the rigid flagella of bacteria. In some organisms, they wave back and forth, staying in one plane. In others, they bend in three-dimensional patterns to generate a helical shape that functions like the rotating helix of bacteria. In some organisms with two-dimensional bending, the flagella are covered with short hairs, called mastigonemes, that are confined to the plane of bending and function to drag more water than a bare flagellum could; mastigonemes reverse the thrust of the flagellum.

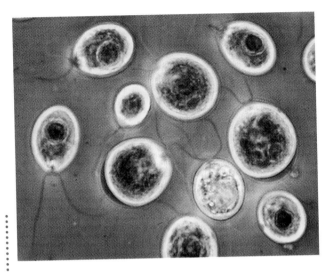
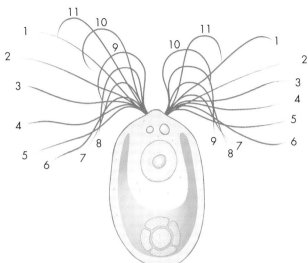

Each of these *Chlamydomonas* cells has one pair of similar flagella that "row" the cell along during swimming. The tracing at right shows how the two flagella of *Chlamydomonas* move to "row" the cell forward. The positions of the flagella were traced from frames of a high-speed movie, filmed at 500 frames per second. On the power stroke (1–6), the flagella are straight, dragging the most water, and move backward. On the return stroke (7–11), the flagella bend to minimize their drag while moving forward. The net result is that water is pushed backward, and the cell moves forward. This photosynthetic green alga is about 9 μm in length.

motion resembling a wave. Initially it was assumed that the basic shape of the beating cilium or flagellum was that of a sine wave, but more careful analysis by Charles Brokaw of the California Institute of Technology indicates that the shapes do not form a smooth curve. Rather, curves of constant radius propagate along the flagellum separated by relatively straight sections. As the wave moves along the length of a flagellum, it carries the surrounding water along with it and produces thrust in a manner similar to the rotating helix of bacteria. In most cases, the waves propagate away from the cell and push the cell away from the flagellum. Sometimes, however, waves propagate in the opposite direction and pull the cell toward the flagellum. Either method seems to work well for locomo-

tion. In many microbes, flagella often move in more complex patterns.

Chlamydomonas is a single-celled green alga that is easy to grow in the laboratory and is a popular experimental organism. This 10-μm-long cell has a pair of 12-μm-long flagella, which do not usually bend in a wavelike motion like those described in the preceding paragraph, but instead beat to the side of the cell in a style more like the breaststroke. During forward swimming, one synchronous beat of the two flagella causes the cell to move ahead 5 μm during the power stroke and backward 2 μm during the return stroke and to rotate through a net angle of 15°. Each flagellum beats at a rate of about 50 strokes per second, about the fastest seen for either cilia or flagella, which may beat as slowly as 10 strokes per second. The cell normally swims forward at a speed near 200 μm/s, a respectable rate for a flagellate, although some can swim twice as fast. The cell can also turn to either side when one flagellum beats more vigorously than the other. And it can swim backward when both flagella propagate waves away from the cell, in the manner described in the preceding paragraph.

Some cells have flagella with hundreds of stiff hairlike protrusions called mastigonemes. They project from opposite sides of the flagellum in the plane of the flagellar beat and act like tiny oars, moving back and forth as bending waves propagate along the flagellum. Mastigonemes can grip much more water than a bare flagellum because of their greater total length. In the well-studied *Ochromonas malhamensis*, a single-celled alga colored orange by its photosynthetic pigments,

the flagellum is 20 μm long, but the total length of the 320 mastigonemes is 350 μm.

The situation is complicated, however, by the fact that when the flagellum bends, the mastigonemes projecting to the outside move through a greater distance and in the opposite direction from those projecting to the inside of the curve. The net result is that those projecting to the outside have more influence, and net thrust is in the same direction as wave propagation along the flagellum, the reverse of the direction of thrust when mastigonemes are absent.

The eukaryotic microorganism *Euglena*, often classified as a green alga, has long posed a challenge to those who would classify all organisms as either plant or animal: it moves around, and can even feed like an animal, but relies mostly on photosynthesis like a plant. It has one long (≈ 100 μm) flagellum that func-

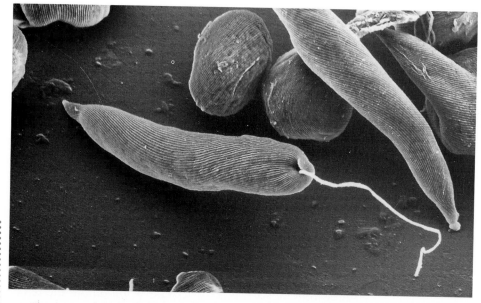

This green alga, *Euglena*, swims by means of a single flagellum, which attaches at the front of the cell. Sometimes it extends forward as seen here, and sometimes it bends around to the side of the cell as seen in the drawing on page 78.

tions in locomotion and a short stub of a flagellum that may simply be a developmental artifact. The functional flagellum carries a single row of mastigonemes 2 to 4 μm long and about 0.1 μm apart. The wave propagates toward the tip of the flagellum, which is normally bent back parallel to the cell. Consequently, the cell moves in the direction of the end of the cell where the flagellum is inserted. All the while the cell is rotating so that the path it takes is actually a helix.

Those protozoa that are covered with cilia and called ciliates can speed past their flagellated cousins, typically moving at speeds near 1000 μm/s—relatively independent of cell size. Ciliates such as the well-known paramecia can be quite large, approaching one millimeter in length. (If they were any larger, they would sink faster than they could swim.) The large ciliate *Paramecium caudatum*, which hunts freshwater ponds in search of bacteria and other minute organisms to feed on, can swim at a rate of 2700 μm/s, or 2.7 mm in a second.

Cilia beat in a very different way from most flagella, moving back and forth with a very asymmetrical stroke. Their action is similar to the breaststroke used by *Chlamydomonas*. In one direction, the power stroke, the cilium is relatively straight and moves perpendicular to its axis. During the power stroke, the cilium sweeps through the maximal thickness of liquid, maximizing drag. During the return stroke, the cilium bends first at the base, as in the power stroke, but the bend then propagates along the cilium so that most of the cilium is kept close to the cell surface. The cilium moves primarily parallel to its length and "drags" relatively little liquid on the return stroke. This asymmetric cycle exerts a net force on the liquid surrounding the cell in the direction of the power stroke, propelling the cell in the opposite direction. In some cells, cilia confine their beat to a plane, but in others the return stroke is off to the side, which keeps the

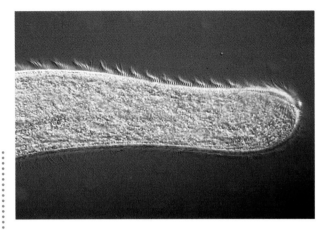

In ciliated cells, cilia adjacent to one another tend to move in a closely coordinated fashion, but the relative phase of the beat cycle changes in a regular fashion over the surface of the cell. This generates visible patterns (called metachronal waves) that appear to move across the cell surface. Here we see a snapshot of the metachronal waves at the top of this large *Spirostomum*.

cilia even closer to the cell's surface. Cells that swim by means of cilia can change the direction of locomotion by changing the direction of the power stroke.

Most ciliated cells have many cilia packed close together, and adjacent cilia beat in close synchrony with one another, since a beating cilium tends to drag its neighbors along with it. Flagella behave the same way: the flagella of different cells will beat together if the flagella are close together. In spite of the high degree of synchrony, neighboring cilia are not exactly in phase with one another, and one can see the phase of beating vary continuously across a broad array of cilia, as if regular waves of activity were sweeping across the array. These "metachronal waves" attract an observer's attention, but their function, if any, is not clear.

It's hard to imagine that a simple rotating motor attached to one end, as in bacteria, could cause the complicated wave patterns of cilia and flagella. Rather, to find the "arm" or motor powering these microscopic oars, we have to look *inside* the beating filaments. The idea that the force for bending is generated within the flagella or cilia is suggested by the dramatic observation that flagella can be detached from a cell, and, if provided with a source of energy in the form of the molecule ATP, they swim.

The interior of both cilia and flagella are packed with microtubules, rigid cylinders about a tenth the thickness of the cilium or flagellum itself. Found in all

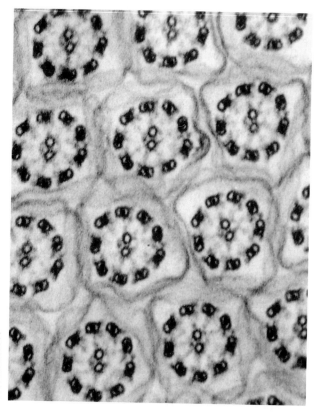

An electron micrograph of a thin section through a bundle of cilia. Here we can see evidence that each cilium contains a circle of nine fused microtubules and a pair of microtubules in the center.

eukaryotic cells, microtubules apparently act as support structures. In cilia and flagella, the microtubules are fused together to form pairs, and crawling down these pairs is the motor that powers bending—a protein called dynein. Each dynein protein is rigidly connected to one pair of microtubules and crawls along an adjacent pair when exposed to ATP—causing one pair to slide along the other. Typical cilia and flagella have a circle of nine microtubule pairs surrounding a central pair running down the axis of the structure, all connected by proteins. When dynein causes sliding on one side of the filaments, the structure is forced to bend.

We can now see how the alternating circular arcs and straight segments of the wave shape might be generated. If microtubule pairs are connected by flexible components, sliding in one region will produce bending in opposite directions at opposite ends of the region of sliding. There is evidence that bending in turn tends to induce further sliding. To explain the straight segments, we can further hypothesize that the sliding activity has a limited time span and is followed by a period during which sliding is inhibited. We now have a model that could explain the basic mechanism by which bending waves propagate along a cilium or flagellum. Somehow the cell must extensively regulate where and when sliding takes place—how we don't know, but the need for such regulation may explain why altogether there are about 200 kinds of proteins present in cilia and flagella.

Feeding with Flagella

Some protozoa and rotifers, particularly those that remain attached to vegetation or even other organisms, use their cilia or flagella to generate currents that carry food particles to the microbes. Cilia and flagella are so thin that they are difficult to see in the light micro-

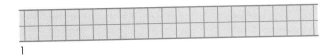

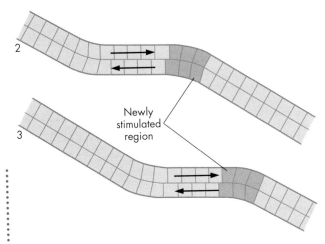

Newly stimulated region

How the sliding of microtubules could generate bending of eukaryotic flagella or cilia. If two long, stiff structures (such as microtubules) are connected along most of their length (1) but slide with respect to one another in one section, bending in opposite directions will be produced at opposite ends of the section (2). Bending in turn may induce further sliding (3). Researchers have clearly demonstrated that microtubules can generate forces that cause them to slide past one another, but they do not yet know how sliding is regulated to generate the complicated movements observed.

cells, one about 20 μm long and lined with two rows of mastigonemes and the other only a third as long and smooth.

During feeding, the longer flagellum beats in a regular pattern, drawing water toward the cell, while the shorter flagellum remains inactive. When the longer flagellum contacts a solid particle, it immediately stops beating and bends to push the particle between itself and the shorter flagellum. Both flagella then seem to move in a way that rotates the particle, as if to assess its desirability. After a second or two, the particle is either released or engulfed.

To release the particle, the shorter flagellum makes a single bend, away from the particle, of 90° or more midway along its length. The longer flagellum resumes its normal beating, and the currents it generates quickly carry the particle away from the cell.

If the particle seems to be edible, the flagella hold it in place while the cell membrane extends to form a "feeding cup" that envelops the food particle within a few seconds, forming a vacuole—a membrane-bound space enclosing the food particle within the cell. The vacuole is retracted back to the cell in about 10 seconds and remains near the base of the flagella for several minutes, apparently while digestion takes place. Eventually, the vacuole is emptied, expelling undigested material, which is carried away by the feeding current.

Considering that cilia and flagella are among the simplest of all cell organelles, this complicated behavior is remarkable. We clearly have much more to learn about even these simple structures.

Surface Tension Supports

At the boundary between water and air is the water's surface, an expanse over which only a few creatures in the macro world are able to move. Water striders can

scope, and they move so fast that it is difficult to observe the details of what happens when a food particle is captured. By enhancing low-contrast images electronically and recording them on tape for playback at slower speed, investigators have recently been able to study the feeding behavior of these microbes.

Cells of the pigmented alga *Epipyxis pulchra* attach at one end to submerged vegetation or larger planktonic organisms in freshwater ponds or lakes. Two flagella protrude from the other end of these walk on water, but most organisms their size and larger would sink if they tried. Microbes seem light enough to stay afloat, but whether they do or not depends on the special properties of a liquid at its surface.

Molecules form liquids because of attractive forces between the molecules; without these forces, the molecules would stay farther apart and form a gas. In the bulk of a liquid, these attractive forces act in all directions on the average, and no molecule feels a net attraction in any direction. At the surface, however, the attractive forces are limited to the liquid side, and this imbalance results in the phenomenon of "surface tension." The liquid tends to move so as to minimize its surface area, subject to the constraint that its total volume must remain constant. In the absence of other forces, a drop of liquid will form a sphere, which is the shape that has the minimum surface for a given volume.

Water molecules have a particularly strong attraction to one another, and water exhibits a surface tension high enough to support a steel needle (as you can easily demonstrate for yourself). Still, you cannot walk on water; its ability to support an object depends on the object's size. Let us take a quantitative look at the ability of surface tension to support an object against the pull of gravity.

Consider a needle or cylindrical organism floating on water, as illustrated in the figure on this page. If the object is not wetted by the water, it pushes down on the surface, stretching it. The increase in surface area is resisted by the surface tension, which acts at an angle to the vertical. In order for the object to remain floating, the vertical component of the surface tension must balance the force of gravity tending to pull the object into the liquid. Our question is: How large can the object grow and still be supported by surface tension?

A simple calculation predicts that for an organism (which will have a density close to that of water) the limit is a radius of about 2 mm, which is consistent with common experience. For example, this is about the body size of a water strider. Thus, even insects can walk on water if they are small enough. Certainly all microbes are sufficiently small to be supported by surface tension.

To use surface tension to float on the surface of water, an organism needs to keep its exterior surface

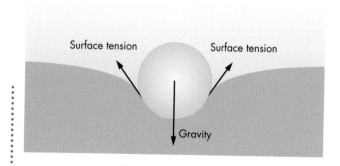 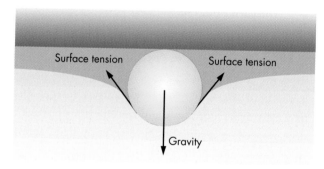

Left: Even an object denser than water can float on the surface of water, if it stays dry and is small enough for surface tension to counteract the pull of gravity. Right: Surface tension can also hold small objects against a solid surface if a thin film of water wets the object and the surface.

Nematodes or round worms, such as this *Caenorhabditis elegans,* are formed of several hundred highly differentiated cells and have well-defined organ systems, including nerves and muscles. Digestion is internal, with food passing in only one direction through a long tube, nearly the length of the animal. Nematodes are extremely common, inhabiting soils and sediments everywhere. Most feed on other microbes, but some are parasites on large plants or animals.

dry—otherwise, the surface would creep over the organism leaving it under water and without support. But most microbes would dry out completely if their surface were dry (because water from inside the organism would rapidly diffuse to the dryer area outside). Therefore, despite their small size, microbes do not often support themselves on top of a pool of water. However, microbes often use surface tension to remain in a thin film of water. A good example is provided by a microorganism whose mechanism of locomotion is very different from any that we have seen so far—the nematode.

Body Undulations

Snakes and eels move without having special appendages by undulating their long slender bodies in patterns similar to flagella. Their sinuous body patterns propagate from head to tail and push the animal forward. Most microorganisms cannot use this strategy, which seems to require distinct muscles and an organized nervous system that gives a high level of control over body bending. All species of nematodes, however, have these abilities and make use of this mechanism of locomotion.

In order for undulatory propulsion to function, the environment must resist lateral movement of the body more than it resists forward movement along the axis. This requirement is the fundamental reason that animals must be long and narrow in order to use un-dulations for locomotion. Nematodes, for example, have bodies that are smooth cylinders about 20 times longer than they are thick.

A peculiarity of nematodes is that they cannot contract body muscles on one side separately from the other side, and consequently they cannot bend their bodies from side to side, like a snake, but must bend toward their upper side or lower side (dorsally or ventrally). To undulate like a snake, then, when on a flat surface they lie on one side or the other. Without a rigid skeleton to strain against, the muscles work against the so-called hydrostatic skeleton provided by the tendency of internal pressure to straighten out the cylindrical cuticle that forms the nematode's outer layer.

In a simple liquid, a cylinder moving laterally meets only about twice the resistance as a cylinder moving longitudinally. Consequently, undulatory propulsion through water is accompanied by a great deal of slip, and nematodes progress through water at a rate that is only about 20 percent of the rate at which the waves of body shape propagate along the body. However, nematodes usually live in environments that provide for much less slipping. Those living on the bottom of bodies of water move through sediments saturated with water. If the pores between particles are sufficiently large, the organisms can snake their way between particles by pushing on the particles without moving them. In sediments with smaller pores, the head of the nematode can act as a wedge and make a small opening for the body to pass through. Because pushing forward requires much less force than moving the body laterally, the nematodes can move without significant slip.

In soil with air spaces, nematodes are held to the surface of soil particles by the surface tension of thin films of water surrounding these particles. As long as the film of water is thinner than the diameter of the

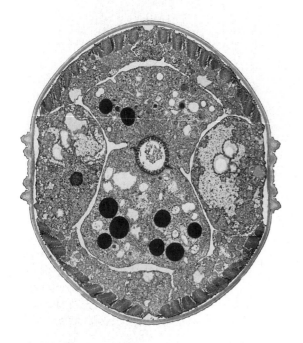

This electron micrograph of a thin section through the center of a nematode shows some of its internal structure. Its digestive tract is in the center and muscles are located above (dorsal side) and below (ventral side) just inside the tough cuticle that surrounds the nematode. Note the treads that roughen the surface of the cuticle on its left and right sides. When the surface tension of a thin film of water holds the nematode to a solid object, the nematode lies on one side (left or right), and the treads on that side grip the surface. The nematode lies on its side because its muscles only bend the body toward its dorsal (upper) or ventral (lower) sides—not sideways—and the surface tension flops it over on one side or the other.

nematode, surface tension presses the nematode to the particle, causing the animal to meet frictional resistance as it moves across the surface. If the resistance to lateral motion is greater than to longitudinal, the nematode can move forward. In some nematodes, this difference in frictional resistance is enhanced by "treads" on the surface of the body. For the small nematodes found in soil (whose radius is less than 20 μm), the surface tension of a thin film of water is thousands of times greater than the force of gravity. For them gravity is meaningless, and they can even move across the bottom surface of an object.

Surface tension endows some nematodes with an ability shared by no other organisms of their small size—the ability to make a jump through air. During a jump the body of an animal is accelerated by a rapid straightening out of the legs (usually). The inertia of the moving body then carries it through the air for a short distance before gravity brings it back in contact with the surface. The gallop of a horse and the run of a human are really a series of jumps.

The famous jumping abilities of fleas, which are almost microscopic, are used to carry these animals to the host bird or mammal. By stretching and then releasing a mass of body protein that is even more elastic than rubber, some species are able to accelerate to more than 100 times the force of gravity and reach heights 100 times their size. We might expect that some larger microbes, not too much smaller than fleas, would have developed similar jumping locomotion.

Although inertia does not often play a role in the locomotion of microbes, there is one report of nematodes jumping through air from one soil particle to another. While investigating the possibility of using insect-parasitic nematodes *(Neoaplectana)* to control insect pests in Australia, E. M. Reed and H. R. Wallace observed that under certain moisture conditions one species could turn itself into a spring. Using

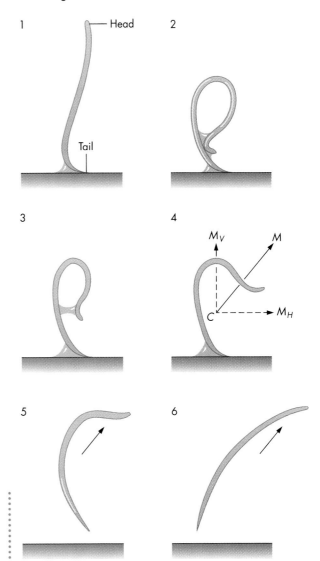

Researchers have observed one nematode species, an insect parasite of genus Neoaplectana, leaping through the air from one soil particle to another. The nematode employs the surface tension of a drop of water to help stretch springlike components of its body. When the surface tension suddenly breaks (3–4), the front of the nematode is accelerated in a near vertical direction, and achieves enough momentum to break the tail's contact with the surface and carry the whole body through the air for a short distance.

surface tension to achieve abrupt release, it could launch itself through the air a distance of several millimeters. The force for straightening the body, still uncertain, might be generated by muscles or by osmotic pressure working against the cylindrical cuticle. The elasticity of the cuticle probably provided an important store of energy.

Creeping and Gliding

The "oars" of protozoa and the "propellers" of bacteria are well adapted for swimming through open water, but cells living on surfaces often employ other forms of locomotion. Many types of cells are able to creep slowly across a solid surface, even though they have no legs and do not move by body undulations. One of the most interesting of the diverse, and sometimes mysterious, mechanisms employed is used by amoebae, which turn their entire bodies into organs of locomotion.

Most amoebae, which lack a cell wall that would confine their bodies in a rigid shape, creep along by extending bulges of the cell called pseudopodia ("false feet") and retracting others. The locomotion of amoebae began to fascinate biologists a hundred years ago, and many (including H. S. Jennings around 1900) made detailed descriptions enhanced by careful analyses. Their most striking observation was that the cytoplasm could be seen to flow from one part of the cell to another—most dramatically from the trailing edge of pseudopodia to the advancing edge. A few types of amoebae were somehow able to move without pseudopodia, although their cytoplasm, too, could be seen to flow—from the back of the whole cell to the front.

The early observers were also fascinated to note that as amoebae creep along, debris particles become

attached to their surface and actually move around, as though carried by the movement of the surface itself. The patterns of movement traced by the particles were especially revealing in the case of amoebae lacking pseudopodia. They demonstrated that the cell membrane rotated around the cell, moving forward along the top of the cell, and backward along its bottom, although the bottom membrane remained stationary with respect to the surface across which the amoeba traveled. Thus, the membrane moved much like the track of a caterpillar tractor.

This type of flow pattern appears efficient for a cell moving across a surface. However, many amoebae probably live in the interstices between soil or other particles, where all sides of the cell would be in contact

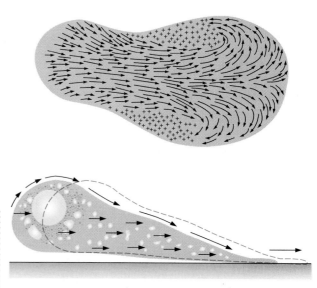

Amoebae are propelled by the flow of cytoplasm. The cytoplasm flows from the rear of the cell into advancing parts of the cell, as seen from the top in the upper part of the figure. In the figure's lower part, a view from the side shows the cell's attachment to the surface and the flow of cytoplasm, indicated by the arrows, that will carry it to the later position outlined by the red curve.

with substrate. This pattern of flow would then simply cause the cell to rotate. Amoebae that extend and retract pseudopodia, with the flow of cytoplasm confined to the center of each pseudopodium, are probably better adapted to life in confined spaces.

H. S. Jennings made some interesting observations of amoeba chasing prey. As an amoeba creeps along, it may encounter a food particle. Whereas in some species food particles tend to stick to the surface of the amoebae, and are immediately engulfed, other species of amoebae are not so sticky, and a creeping amoeba tends to push an unattached food particle ahead of it. For example, Jennings observed an *Amoeba proteus* attempting to engulf a spherical *Euglena* cyst. As it pushed against the cyst, the cyst repeatedly rolled away, as Jennings illustrated in the drawing reproduced on this page. Most remarkably, the amoeba continued to pursue the cyst for more than 10 minutes without ever capturing it.

In another case, an *Amoeba proteus* extended pseudopodia along both sides and over the top of a *Euglena* cyst it had encountered. As the pseudopodia extended past the cyst they bent toward it and formed a pouch that prevented the cyst from moving away. Eventually the pseudopodia met on the opposite side of the cyst and fused. Within two minutes of the first encounter, the cyst was enclosed in a vacuole inside the amoeba.

Since these early observations were made, about a hundred years ago, the main question has been: How are the flows of cytoplasm generated? Many hypotheses have been suggested, but a clear understanding remains to be obtained. The most common suggestion is that the flows are generated by the cytoplasm changing back and forth from solid to liquid as fibers of a protein called actin undergo changes in length and cross-linking. Most of the cytoplasm near the surface and in the rear of the cell is thought to be in a

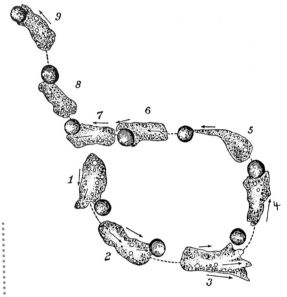

An *Amoeba proteus* attempts to engulf an encysted *Euglena*. But the cyst is easily pushed away, and the amoeba has great difficulty wrapping its pseudopods around its prey. Nonetheless, it demonstrates remarkable persistence.

"gel" state; here the actin fibers are long and frequently cross-linked. In the center of the cell and in advancing pseudopodia, the cytoplasm is in a fluid "sol" state with shorter, less cross-linked actin fibers. A recent model proposes that the following processes take place. Gelled cytoplasm approaching the trailing edge of the cell experiences an increase in calcium (Ca^{2+}) concentration, causing fibers to slide and contract, as in muscle. The contraction squeezes fluid out of the gel, and the fluid flows forward along the top or through the center of the cell. Eventually, the actin fibers at the rear of the cell break down, and the constituent molecules are carried forward by the extruded fluid. At the leading edge, these molecules reassemble under conditions of low calcium concentration to form gelled cytoplasm again. As locomotion continues, these actin fibers are once again found at the rear of the cell.

Cells often simultaneously produce multiple pseudopodia that seem to move independently of one another and even to be in competition. Eventually all but one of the pseudopodia contract, and the cell contents flow into the remaining pseudopodium. This process of extension and contraction proceeds continuously, creating the forward motion of the traveling cell. Remarkably, a cell can be cut in two and each half will move on independently of the other.

Many individual cells of multicellular animals move in a fairly similar fashion, particularly certain immune system cells that defend our bodies against pathogens, and cells finding their proper locations in the developing embryo. The locomotion of these animal cells has been studied more thoroughly than the locomotion of free-living amoebae. Particles attached to the top surface of such an animal cell move away from the advancing edge of the cell, suggesting that membrane is produced at the leading edge and flows toward the back of the cell, as expected for cells moving through a three-dimensional matrix.

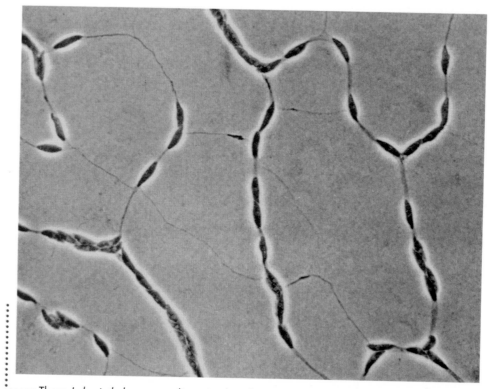

These *Labyrinthula* are peculiar microbes that spend most of their life cycle living in a network of tunnels as seen here. The cells move rapidly back and forth along the tunnels, but outside the tunnels they cannot move at all.

Diatoms, such as the two shown attached to seaweed in this scanning electron micrograph, live wherever there is water and sunlight. The elongated one clearly shows the slitlike opening called the raphe that extends along the long axis in some diatoms. Diatoms that contain this structure are capable of gliding locomotion.

The marine slime mold *Labyrinthula*, which parasitizes eel grass, has a unique approach to motility. To the casual observer, the slime mold looks like a slimy mass on the surface of the grass, but under the microscope the observer sees a colony of spindle-shaped cells that travel together in narrow streams. The cells shuttle around within a membrane-bound "slime way" that they themselves secrete. This slime way contains the muscle proteins actin and myosin. Apparently the cells modify the calcium concentration in the slime way to regulate the sliding of the muscle proteins, and this sliding action somehow causes the cells to move.

A wide variety of single-celled organisms with rigid cell walls creep along without changing cell shape, leaving behind a trail of slime, using a method of locomotion called gliding. The best-known of these organisms among the prokaryotes are photosynthetic cyanobacteria (also known as blue-green algae), the social myxobacteria, and the extremely small mycoplasmas, which are known for lacking a cell wall. Among eukaryotes, gliding is the method of locomotion used by a variety of protozoa. Gliding is such a slow means of travel that even some amoebae could beat gliders in a race. Bacterial cells, for example, have been observed gliding at speeds of 0.04 to 11 μm/s. The fastest of these gliders would take over a minute to travel a millimeter, while the slower ones could take hours.

From the diversity of gliding organisms it is likely that there are several mechanisms of gliding, but they remain mysterious. Scientists have made "snapshots" of gliders in motion using interference reflection microscopy, which is sensitive to differences in the separation of a cell from a glass surface in the very small range of 10 to 100 nanometers (a nanometer, abbreviated nm, is a billionth of a meter). They have discov-

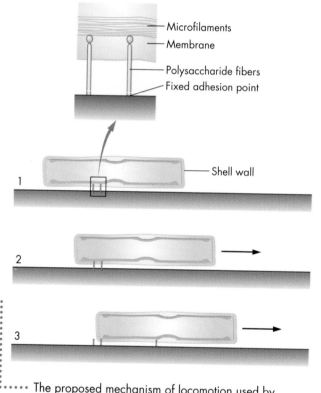

The proposed mechanism of locomotion used by diatoms having a slit between the two halves of their cell wall. Polysaccharide fibers are supposed to extend through the slit; at one end the fibers are attached to the surface underneath and at the other end to membrane proteins. The membrane proteins slide along microfilaments inside the cell, pulling on the fibers. Since the fibers are anchored to the surface, the microfilaments move, pulling the cell across the surface.

ered that even the parts of cells in close proximity to the surface can glide, a finding that eliminates as possible mechanisms the kind of creeping used by amoebae as well as the looping action of inchworms. In eukaryotes, gliding may be based on movements of individual molecules rather than parts of cells. For example, a protein sliding along a microtubule and attached to the surface across which the organism glides could help propel a cell.

Perhaps it is most surprising to find gliding behavior in diatoms, which have a pair of hard shells made of crystallized silica that fit together like a pill box and its lid and cannot change shape. They move at rates measured at 0.2 to 25 μm/s. The gliding locomotion of diatoms has intrigued biologists for two hundred years, and many different mechanisms of propulsion were proposed in the nineteenth century. Some would have made use of jets of fluid expelled by the diatom, but jets depend on inertia for propulsive force and would not function in the viscous world of microbes.

Microbiologists lack detailed evidence of the mechanism by which diatoms propel themselves, yet by putting to use the latest techniques for electron microscopy and their knowledge of how other types of cells move they have been able to put forward a solid hypothesis. All the motile diatoms have a pair of narrow slits, each less than 0.1 μm wide, in at least one shell. In the cytoplasm behind the slits are bundles of microfilaments composed of the muscle protein actin. Motion is probably generated by proteins sliding along these filaments, and the movement is mechanically linked to the surface by polysaccharide fibers attached to the proteins and extending through the slits.

The speed at which microbes propel themselves may seem slow—from a few micrometers to a millimeter per second. Yet if one considers the speed relative to body size, there is a remarkable constancy, illustrated in the plot on the facing page. Even including

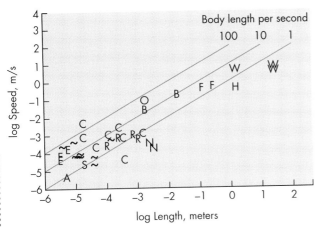

The speeds of motile organisms vary more than a millionfold—from a few micrometers per second for some bacteria up to several meters per second for whales. But size changes even more. If we compare speed with size, there is clearly a correlation—larger organisms are faster. As a rule, most organisms seem able to achieve a maximum sustained speed of between one and a hundred body lengths per second no matter what their size. Viewed in this way, relative to their size, some microbes are among the fastest of all organisms. Each symbol indicates an observation of maximum sustained swimming speed for a species belonging to: archaebacteria (A), spirochetes (S), flagellated eubacteria (E), flagellated eukaryotes (~), ciliated eukaryotes (C), rotifers (R), nematodes (N), ostracodes (O), water beetles (B), fish (F), whales (W), humans (H).

the largest animals, the fastest sustainable speeds depart little from the range of 1 to 10 body lengths per second. Thus, relative speed is nearly constant in spite of millionfold changes in size and absolute speed. There must be some basic physical constraints at work here in spite of the wide range (10^{13}-fold) of Reynolds numbers involved. Remarkably, if judged by their relative swimming speed, bacteria are among the fastest of all organisms.

The physics of the micro world requires that bacteria swim fast to accomplish anything. Ideally, a bacterial cell would swim fast enough to increase the rate at which nutrients diffuse to it, but a sphere of radius 1 μm would have to achieve a speed of 700 μm/s to increase its intake of nutrient molecules by only 10 percent, and such high speeds may be impossible. However, a bacterium can swim far enough before losing orientation to sample the concentration of nutrients in a new location; it can then use this information to navigate toward the nutrients' source. Even this possibility would be closed to a slower-swimming bacterium because the diffusion of small molecules would effectively keep it in the same chemical environment. Thus, small cells must either swim very fast or there is little point in swimming at all, and the pressures of natural selection probably caused small bacteria to evolve to swim fast—relatively.

The myxobacterium *Chondromyces crocatus* forms this multicellular treelike structure, only half a millimeter high, to lift the spores at the tips of the branches up away from the substrate. In this position the spores are more likely to be dispersed to distant

Catching a Ride

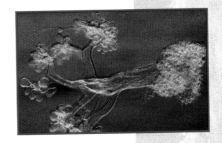

As some bacteria grow and divide into daughter cells over many cycles, the progeny of an initial cell accumulate in a densely packed colony. Growth eventually slows because the dense population of bacteria depletes nutrients faster than they diffuse into the colony. Microbiologists make use of this behavior in counting and characterizing many kinds of microbe by the physical appearance of visible colonies formed on Petri dishes filled with an agar gel and nutrients. In nature, however, the colony probably faces extinction unless its members disperse to new habitats offering fresh sources of nutrients.

All organisms face this same challenge of getting their progeny dispersed through the environment and separated so that they do not compete with one another. Even among humans, adolescents tend to leave home and travel. But many organisms, especially plants, are attached to one spot and cannot move away. These immobile organisms often produce specialized structures for dispersing their progeny. The small seeds of dandelion plants, for instance, catch air currents in their many long "hairs" and drift far away. Fruits are essentially bribes to animals to carry seeds away, while the "beggars lice" of other plants stick to mobile animals for dispersal. Tumbleweeds roll across the prairie pushed by winds and dropping seeds as they go.

Fungi and other immobile microbes face the same challenges as plants and employ similar strategies. Many of these microbes produce small structures, called spores, that function as seeds to disperse progeny through space. Spores usually are resistant to environmental stresses such as drying and heat and can survive long periods. Some must pass through periods of dormancy before they will germinate and, in a sense, are dispersed through time. Although many spores are easily carried by wind or water currents, or even animals, getting the spores away from the parent

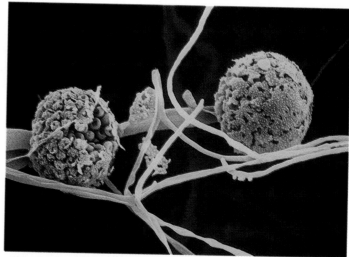

Bread provides a rich medium for the growth of a variety of fungi (left). The strands of hyphae of which fungi are formed are clearly visible in a colorized scanning electron micrograph (right) of the mold *Mucor mucedo*, growing on a piece of bread. In the center of the micrograph may be seen two sacks (sporangia) containing spores. The sporangium on the left has split open, and spores are visible inside.

Typical small fruiting bodies of a fungus, *Hemitrichia calycuta*. The spores are formed and released from the yellow spheres, which are lifted above the surface on stalks.

is initially a tricky problem in the special conditions of the micro world.

Riding the Wind

A microbe that simply released spores from its surface wouldn't be too successful at dispersing them, even if wind or water currents were present to carry them off. The reason has to do with the dominance of viscosity at small scales.

Even in turbulent flows, the effects of viscosity are so strong that vortices below a certain size do not exist. In the natural environments of ocean and atmosphere, the smallest vortex size is on the order of 1 cm. The existence of a limiting vortex size means that flow is always laminar sufficiently close to a surface, in a region called the laminar or viscous sublayer.

No stirring occurs within this layer, and the only mechanism of transport through it is diffusion. For leaves, the unstirred layer normally ranges in thickness from 0.1 mm (at high wind speeds) to 1 mm (at low wind speeds). Microbes are almost always surrounded by a laminar flow layer that is thicker than they are.

A microbe's main problem in getting its spores dispersed by wind or water currents is that the spores tend to become trapped in the laminar sublayer of fluid surrounding any surface and flowing parallel to it. Spores released right at a surface would, at most, simply roll along the surface pushed by the wind or current until caught on a protrusion. You may have noticed how dust can stick to a fan blade; it is very difficult to blow a fine powder from a solid surface because the flow is always slow close to the surface and parallel to it. If spores are released above a solid surface but within the laminar sublayer, the most currents can do is carry spores along until they fall back under the force of gravity. However, if the spore can be released above the laminar sublayer, in turbulent flow, then random currents will sometimes carry the spore away from the surface and transport it long distances to a completely separate surface.

A common solution to getting out of the boundary layer is for a microbe to release spores from a stalk that sticks up from the surrounding surface. Since the laminar sublayer is generally only a few millimeters thick, the stalk does not have to be very tall. This strategy is used by a few bacteria, some slime molds, and many fungi.

Primitive fungi are mostly composed of slender filaments called hyphae that lack partitions between nuclei to create separate cells, although more "advanced" fungi form cell-wall partitions at intervals along the filaments. When conditions are favorable, hyphae can grow into a large network called a mycelium. Fungi usually live unobserved, with their mycelium extending within a dark, moist environment such as rotting wood or dung. When they become apparent to the casual observer, it is most often because the mycelium has produced specialized stalks or other spore-bearing structures, called fruiting bodies, that protrude from the surface. Toadstools are large versions of these structures, composed of a mass of closely adhering hyphae; the formation of many small fruiting structures that release spores into the air where winds can carry them long distances makes moldy bread look fuzzy.

Only one group of prokaryotes produce fruiting bodies with spores on stalks. The myxobacteria are gliding bacteria often found on rotting wood. They live in swarms of cells that secrete digestive enzymes to break down food externally. When food becomes scarce or conditions otherwise deteriorate, the mass of cells reorganizes to form a stalked fruiting body. Since

Even some bacteria form stalks to lift their spores away from the surface. This fruiting body was formed by myxobacteria belonging to genus *Chondromyces*.

the fruiting body only projects about 10 μm from the surface, it is not really out of the boundary layer, and it is not clear what the main agent of dispersal is.

Gravity and Dispersal

Once a spore is caught in a current of air or water, how far it travels before coming to rest is influenced by the same property of viscosity that makes a car run less efficiently at higher speeds: viscosity offers more resistance to movement as speed increases, and at higher speeds more force is needed to overcome it. If the force acting on an object in a viscous fluid remains constant, the object will accelerate to a speed at which the force is balanced by the resistance to movement caused by viscosity. This speed is called the terminal velocity. In nature, the most common constant force is that of gravity, and viscosity influences the rate at which raindrops fall through air and bubbles rise through water.

Gravity also pulls on microbes and spores as they travel in a current, dragging them toward the surface beneath. How fast would a microbe fall in air or sink in water? The force of gravity acting on a particle in a vacuum is equal to its mass times the acceleration of gravity ($g = 980$ cm/s^2). Microbes do not live in a vacuum, of course, and so we must take into account the surrounding water or air. If a particle is immersed in a medium of density nearly as great as that of the particle, then buoyancy will cause the particle to sink only very slowly or not at all. In this case, the net force acting on the particle is much reduced, but can still be calculated as mass times the acceleration of gravity by replacing the mass of the particle in the calculation by its "effective" mass. The effective mass is

the particle's mass reduced by the mass of the displaced medium, which is equal to the volume of the particle times the density of the medium.

The effective mass of a microbe immersed in some medium is easily calculated if we assume a spherical shape for most microbes, and express the mass of the sphere as its density times its volume. The effective mass is simply the sphere's volume ($\frac{4}{3}\pi r^3$) times the difference in density between the sphere and the medium. This force of gravity (acting to accelerate the sphere) is balanced by the force of viscosity (acting to resist movement of the sphere), which is given by Stoke's law. Equating the two forces and solving for the velocity produces the equation:

$$v_{terminal} = \frac{2r^2}{9\eta}(\rho_{sphere} - \rho_{medium}) \times g$$

It can be seen that the terminal velocity is proportional to the difference in density of the sphere and the medium. The velocity can be either upward or downward according to whether the particle is less dense or more dense than the medium. If the density of the particle matches that of the medium, the terminal velocity is zero, the particle neither sinks nor rises, and the particle is said to be neutrally buoyant. In addition, the terminal velocity is inversely proportional to

Computing Terminal Velocity

Subject (radius, μm)	Medium	Viscosity, g/cm s	Particle density, g/cm³	Medium density, g/cm³	Terminal velocity, cm/s
Bacterium (1)	Air	0.00018	1.2	0.0012	0.01
Spore (10)	Air	0.00018	1.2	0.0012	1.5
Bacterium (1)	Water	0.010	1.2	1.00	0.00004
Rotifer (100)	Water	0.010	1.2	1.00	0.4

viscosity. Although the equation itself applies only to a sphere, these relationships would apply to a particle of any shape.

We can also see that for a sphere the terminal velocity is proportional to the square of its radius. Thus, small things such as microbes have very low terminal velocities: compare the values given in the table on the previous page to the typical speeds of winds (10 to 1000 cm/s) or water currents (1 to 100 cm/s). Microbes and their spores sink at speeds that are generally much lower than the normal speed of wind or water currents, and they can be carried long distances before they reach bottom and stop traveling.

The spores of fungi, with diameters in the range of 5 to 50 μm, have sinking rates in air of 2 or 3 cm/s for the larger ones down to 0.05 cm/s for the smaller ones. Since wind speeds are typically more than 10 cm/s, these fungal spores have low enough sinking rates to be easily carried by winds.

Breaking Free of Surface Tension

Microbes dependent on the wind to disperse their spores face another basic problem, in addition to escaping from the laminar sublayer. Even a spore released from a stalk tends to be held to the stalk by surface tension. Toadstools, which are the largest stalks

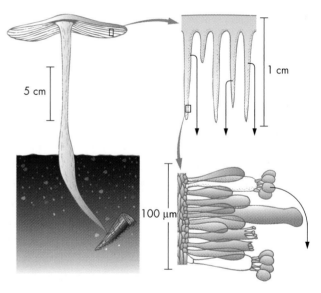

A mushroom's gills are clearly visible to anyone looking upward at the bottom side of its cap (left). Spores are released from the gills in such a way as to become airborne (right). The gills are lined with tiny projecting structures called basidia (lower right), at the tips of which develop the round spores. The fungus flings its spores horizontally from the surface of the gills to break them free of surface tension. Following the path indicated by the black arrows, they fall out between adjacent gills into the open air below the cap, where there is a good chance the wind will catch them and carry them to distant environments.

formed by microbes, have a mysterious mechanism for spore release that probably serves to free the spore from surface tension. The underside of the toadstool "cap" is lined with thin, vertical flaps of tissue, called gills, that radiate like spokes from the stalk to the cap's outer edge. Somehow the toadstool is able to shoot the spores a few hundred micrometers away from the gill surface on which they were formed into the air space between the gills. Since the gill surface is oriented vertically under the cap of the toadstool, the spores subsequently fall by the force of gravity out from between the gills into the relatively open air below the cap. They thus get several centimeters away from any surface and have several seconds to be caught by a wind before reaching the ground. Since there is only a narrow space between adjacent gills, the gills must be oriented precisely to the vertical; otherwise, spores would land on another part of a gill. Accurate orientation is achieved by having a stout stalk that does not sway in the wind and gills that are capable of geotropism (that is, they grow in accurate alignment with the direction of gravity).

Some types of fungi succeed in avoiding the surface tension problem altogether. They eliminate the large surface tension present where water meets air by allowing the spore-containing tissues to dry up, although there will still remain the weak attraction that exists between all solids. Some dry-spore fungi seem to rely simply on wind to launch their spores, but more commonly species have special mechanisms to help the spores break free.

For example, some downy mildew fungi develop fruiting structures only when the air is saturated with water. When the air dries, the structures lose water and shrink, but unequally, in such a way that the branched stalks twist. Often the branches become entangled, preventing free rotation and building up

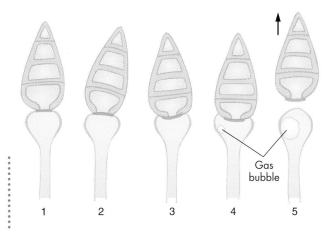

The mechanism used by the fungus *Deightoniella torulosa*, a common pest on banana trees, to propel a cap containing spores. The stalk of the fruiting body loses water through drying, and the thin wall attaching the cap to the head of the stalk becomes indented (1–3). Continued water loss reduces the pressure in the head. Suddenly, a gas bubble forms (4) and rapidly expands, causing the indentation to evert and kicking the cap away (5).

stresses. When the branches do break free, they jerk, releasing their spores.

Other fungi generate jerking motions or kicks by a liquid-to-gas phase change. As the stalk dries out, the water evaporates from an internal water-filled space, and that space decreases in volume. This causes distortions in the walls, which in turn cause the stalk to bend. As evaporation continues, the internal pressure declines until it is below the boiling point of the water solution at the ambient temperature. The pressure can drop well below the boiling point, producing a "superheated" solution. But sooner or later the water abruptly turns to vapor; suddenly a gas bubble appears, which easily expands, releasing tension on the walls and allowing them to return to their original shape. The phase change is so rapid that it causes a jerk or kick. *Deightoniella torulosa,* a fungus that causes spots on banana plants, projects its spores up to 2 cm by this mechanism.

An alternative mechanism for spore release is found in the puffballs formed by some fungi. These fungi keep dry spores in a thin-walled sphere several centimeters in diameter that is pierced by a small hole at the top. When the walls are indented—often by raindrops—air is blown out of the sphere, carrying spores with it. In a sense, then, the puffball is able to generate its own air currents to help launch the spores.

A spore carried off by wind or water has to find a resting place sometime or it will never develop the intimate contact with its environment necessary for nutrient uptake and growth into a mature fungus. Just how a spore is brought to a stop depends on its inertia, and thus on its size. A large spore, with greater inertia, being carried by the wind can't change direction quickly when the wind changes course to flow around an object; it will smack into an obstacle and cling to it. Fungi that live on vegetation usually have larger spores that can be intercepted by plants in the path of the wind. Small spores tend to be carried around objects as the air flows around them. Such spores usually belong to species that live in the soil, and they eventually reach the ground when they settle out or are washed out by rain.

Riding an Animal

Although many fungi release dry spores that are carried by winds, many others produce spores embedded in a layer of liquid slime. Held tightly by surface tension, these spores could never be picked up by winds. How are they dispersed? One answer is that these spores are often designed to be dispersed by animals.

The fungus *Ceratocystis fagacearum* grows under the bark of oak trees, causing a disease called "oak wilt." A mature mycelium swells, splitting the bark, and produces sticky spores, as well as volatile chemicals similar to those given off by fruit. The smells attract sap-feeding beetles, which pick up the sticky spores and spread them to other trees. The beetles may also

A cloud of spores is expelled from puffballs by the impact of raindrops. This fungus, *Calostoma cinnabarina,* was photographed in the Monteverde Cloud Forest Reserve, Costa Rica.

The spores of stinkhorn fungi (such as this *Dictyophora duplicata*, commonly called the netted stinkhorn) are spread by flies—look closely at the cap. As their name suggests, these fungi give off a powerful scent, and it is this scent that attracts the flies. The small, smooth spores are presented in a sugary slime, which the flies ingest.

benefit, because they grow better when feeding on infected trees than on healthy trees, probably because the fungus provides better nutrition.

Other fungal spores *(Tieghemiomyces)* are attached to long sticky threads that may adhere to the whiskers of the mice that produce the dung on which the fungus lives.

Truffles employ a different strategy. They give off a scent that is attractive to rodents, which dig up the underground spore-forming structures and eat them. As the animal moves around, the spores pass through its digestive tract and are deposited unharmed with its dung.

Sometimes the entire microbe finds a convenient ride. Nematodes of many species get themselves transported by insects. For example, several species that live in dung pats routinely acquire transport to fresh dung by riding on dung beetles. Although the behavior of the nematodes that causes them to become attached to the insects is not well understood, a common observation is that the nematodes climb up onto a surface and project their bodies out into space attached only by their tails. They wave around, and if their head end contacts another surface, they are somehow able to break the surface tension holding their tail and become attached to the new surface. If this new surface is an insect, they can now hitch a ride, and there is a good chance that the insect they have attached to is also interested in moving to a fresh dung pat.

Riding a Splash

The splash of a raindrop may provide the means of spreading the spores that some fungal species form in slime. When a fast-moving drop hits a thin layer of liquid, thousands of droplets are formed in the splash, and each droplet contains a mixture of the liquids in the drop and in the layer. Thus, if the layer contains spores, they are picked up and carried in the droplets of the splash.

Let us consider the possibilities. When the raindrop's downward fall is abruptly brought to a stop

Typical raindrops are from 0.5 to 5 mm in diameter and fall at speeds of 500 to 800 cm/s. At these speeds, a raindrop has enough kinetic energy to raise all the water in the drop a distance of 1 to 3 meters. In reality, not all the energy during a splash goes to accelerating the water, but clearly there is great potential for dispersing microbes many centimeters.

Experiments have shown that a single large raindrop without wind can disperse over a thousand spores to different locations. Most spores land more than 20 cm away, and a few land more than 50 cm away. Given the mechanism's simplicity and success, it is probable that many small adult microbes, including bacteria, are often dispersed by splash as well.

Shooting Projectiles

There is a diverse group of fungi that live primarily on the dung of cows and horses and other large mammalian herbivores. These animals ingest the spores of the fungi along with the grass they eat and pass the spores uninjured through the digestive tract; the intestinal environment may even stimulate germination. The fungi then grow in the deposited dung and eventually develop spore-bearing reproductive structures. But since the herbivores avoid eating grass near dung piles, the fungi need to get their spores dispersed onto the grass some distance away.

A behavior that several unrelated fungi have developed is to shoot their spores as projectiles. The fungi shoot off the spores only during daylight in a large, sticky spore mass, which they can propel a meter or more away. The launching apparatus turns to follow the light, so that the projectiles are aimed at the sun or brightest part of the sky. This strategy probably ensures that the projectile is launched at an angle to the vertical, at least when the sun is visible, and is aimed at an opening, if the fungus is covered by debris.

Superimposed photographs show an individual of the insect parasitic nematode *Steinernema carpocapsae* in the infective stage as it tries to catch a ride. The nematode is standing on its tail, held to a splinter of wood (black) by surface tension, and waving in the air. To make the image, I took 10 photographs at one-second intervals, digitized them, and superimposed them, inserting a different hue for each photograph. The color scale at the top indicates the order in which the hues were used.

by its contact with the slime layer, its energy of motion, or kinetic energy, must be expended somehow, and the drop's downward motion becomes the upward motion of the splash. How much kinetic energy ($\frac{1}{2}mv^2$) the drop has depends on its mass and velocity.

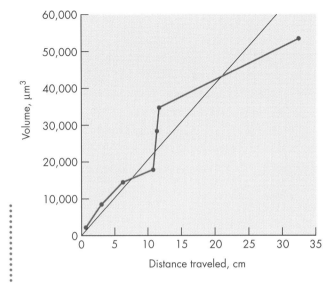

Measurements were taken of the average distance at which spore-carrying projectiles landed for seven species of fungi, then plotted against the total volume of all spores in the projectile—the payload. The measurements reveal a strong correlation between spore projectile size and the distance it is propelled.

As we have seen, many other fungi shoot their spores in order to break surface tension and launch them into the air, but dung-growing species throw their spores much farther—some even land meters away. Altogether, fungi employ a variety of mechanisms to accelerate the projectiles. These include the rapid generation of gas as in guns, the expulsion of jets of water, and dehydration to bend and stretch springy materials, which store energy as in catapults and bows.

A common observation about fungi that shoot spores is that larger spores carry farther, as can be seen in the graph on this page. From what was said in Chapter 2, we might imagine that larger spores have an advantage because the larger an object, the more important inertia is compared to viscosity. Assuming that a large spore and a small spore attain equal veloc-

ity after launch, the large spore's inertia will keep it moving longer. But, you may object, wouldn't it take more energy to accelerate a large spore to the same speed as a small one? Can the organism afford to expend the extra energy? As the box on the facing page explains, even when an organism is thrifty in its energy expenditure, and doesn't launch its spores to high speeds, those spores will still go farther if they are

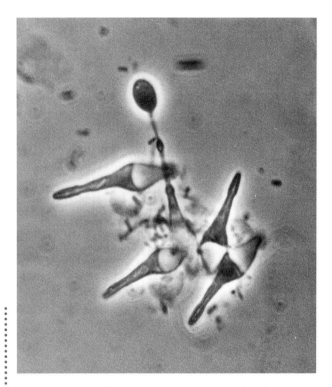

The parasitic fungus *Haptoglossa mirabilis* infects its rotifer host by means of a gunlike injection cell only 8 μm long. A swimming rotifer creates currents that trigger the cell to shoot a tube through the animal's cuticle and then pump a spore through the hypodermic-needle–like device into the rotifer. In this photograph, there are four such cells ready for action. A fifth, with its base in the center of the cluster, has fired but failed to penetrate a host; its spore is the bulb at the top.

Do Large Particles Travel Farther?

If a solid sphere (of density ρ) is propelled to a velocity of v_0 and then allowed to coast in a viscous fluid with negligible density (air), its speed will decrease exponentially:

$$v(t) = v_0 e^{-t/\tau},$$

where the time constant is

$$\tau = \frac{2r^2\rho}{9\eta}$$

The distance coasted at time t is

$$d(t) = v_0 t(1 - e^{-t/\tau})$$

and the distance ultimately coasted is

$$d(\infty) = v_0\tau = \frac{2r^2\rho v_0}{9\eta}$$

Consequently, the distance ultimately coasted is proportional to the square of the size, assuming that

spheres of different sizes are propelled to the same speed. This calculation indicates that larger projectiles have an advantage, if velocity is the limiting factor.

However, it takes more energy to accelerate a larger particle. What if energy is limiting? The minimum energy required to accelerate a particle is $\frac{1}{2}mv^2$. If the particle is a sphere of density ρ, the energy is

$$E = \frac{2}{3}\pi\rho r^3 v^2$$

Solving for the velocity and substituting into the previous equation for distance, one finds that the ultimate distance coasted is proportional to the square root of both energy and radius. Thus even when energy is the limiting factor, larger particles travel farther, at least up to the point at which the particle becomes too large (Reynolds number ≈ 1) and turbulence starts to increase the drag on the particle.

large, as long as they are not so large that they enter the realm where turbulence becomes a factor.

The fungus *Sphaerobolus stellatus,* which also grows on dung, forms a fruiting body only 2 mm in diameter but manages to catapult its sticky spore mass several meters away. The growing fruiting body develops a two-layered hemisphere, forming a cup opening toward the light within which rests the 1-mm-in-diameter spore mass. The inner layer of the cup has a high concentration of glycogen, which is rapidly converted to glucose. This increase in glucose raises the osmotic pressure, causing water to flow into the tissue and generating pressure to expand. Suddenly, the inner layer expands by eversion of the hemisphere so that the inner layer pops out of the "cup" like a bal-

loon. This rapid movement of the layer launches the spore-containing mass.

In *Pilobolus,* the spore mass is propelled by a jet of water squirted out the end of the stalk on which the spore mass sits. The jet of water bounces off the spore mass and imparts momentum to it. Various measurements indicate that the spore is accelerated to a speed of 10 to 20 m/s. At this speed, the Reynolds number is about 300, and turbulence increases drag two- or threefold over what would be predicted by Stoke's law. Nevertheless, the spore mass can be projected a distance of 2 meters or more.

If this isn't sufficiently amazing, nematodes exploit this fungus to obtain a free ride. Adults of the nematode *Dictyocaulus viviparus* live in the lungs of cattle, causing "parasitic bronchitis." Larvae of the nematode develop in dung pats. When they reach the infective third stage, they often crawl toward light up the stalks of *Pilobolus* also growing in the dung. They then coil on top of the spore mass and wait quietly. When the spore mass is discharged, the nematode is propelled along with it away from the dung pat. If lucky, it will land on vegetation that a cow will soon eat.

Controlling Host Behavior

The microbes, often bacteria, that cause infection are of course dreaded for their ability to bring sickness or even death, but the way they spread invisibly from one person to the next is especially frightening. We have seen several examples of a microbe making use of the locomotive ability of an animal, and one example of a nematode that exploits the explosive spore release of a fungus. These examples lead one to ask whether infective microbes can disperse themselves by influencing the behavior of their host. There are really two different questions to ask. The first is: Do infected hosts exhibit specific behaviors that aid in dispersing the infectious microbe? The second question is: Do the microbes cause this behavior for the purpose of dispersal? The first question is relatively easy to answer, while the second is quite difficult.

It is easy to point to examples that answer the first question. We all know that common symptoms of illness are coughing, sneezing, having a runny nose, or suffering from diarrhea. All of these symptoms are potentially effective mechanisms of transmitting microbes to other people. But are they caused by the invading organisms for this purpose, or are they part of the host's protective mechanisms, or are they simply side effects of infection? Here are a few examples to consider.

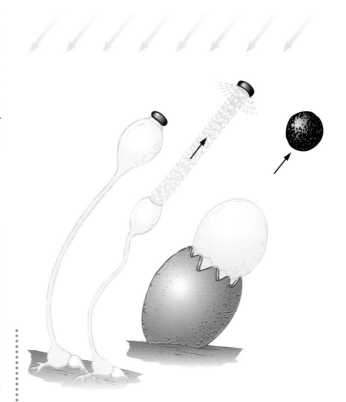

The two fungi illustrated here live on the dung of herbivorous animals and fling their spores toward the light. On the left, *Pilobolus kleinii* bends its stalk to aim toward light. During discharge, a jet of water from the stalk accelerates the spore-carrying projectile, which can be propelled several meters. On the right, *Sphaerobolus stellatus* also aims its mortarlike gun at the light.

Coughing and sneezing are employed by the body to keep the airways clear of obstructions. By blowing out particles and fluids, this behavior not only rids the body of some microbes but also effectively disperses them. In many cases, it would appear that the microbes benefit more in getting passed to another susceptible host than the host benefits from eliminating a small fraction of the microbes in its body. However,

During a sneeze, a cloud of small droplets is expelled with great force. Each of these droplets can carry many bacteria, and the sneeze can very effectively spread microbes through the environment.

we lack the detailed knowledge we would need to determine whether the microbes behave in a way to promote sneezing or coughing.

Cholera is a serious threat to human beings. During a series of outbreaks that ravaged London in 1853 and 1854, John Snow carried out one of the first epidemiological investigations of any disease and demonstrated that cholera was transmitted by drinking water contaminated by sewage. Thirty years later the cause of the disease was identified as a motile bacterium, named *Vibrio cholerae*. After being ingested, the bacteria adhere to the epithelial cells lining the huge surface of the small intestine and multiply. In a healthy person, 5 to 10 liters of water enter the small intestine daily, and 90 percent is normally absorbed and transported to the blood thanks to the activity of the epithelial cells. However, the *V. cholerae* secrete a protein that causes a modification of these cells so that they no longer absorb water efficiently. The result is the defecation of large amounts of water—diarrhea. The consequences are dehydration of the host (which is likely to be lethal) and dispersal of *V. cholerae* (if facilities and human behavior do not confine the excrement). Since the death of the host is of no advantage to the bacteria, and inhibition of water absorption has no apparent advantage other than dispersal, one can tentatively conclude that this behavior serves to disperse *V. cholerae*. Other bacteria probably cause diarrhea for the same reason.

Entering a new host is only one step in the infection process. The ability to attach itself to a surface is often another requirement for a microbe infecting another organism. To enhance attachment, bacteria often form pili—protein filaments 1 to 2 μm long, which cover their surfaces. A good example is the famous bacterium *Escherichia coli*, which is found in the

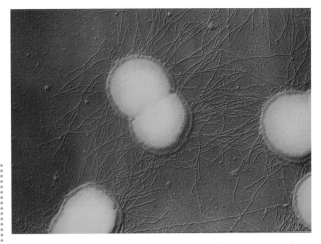

This colorized electron micrograph shows a type of nearly spherical bacteria called cocci. The filaments in the background are pili, which help attach the cells to surfaces.

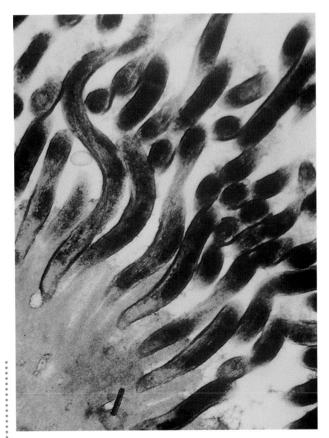

This colorized electron micrograph of a thin section through the surface of the intestine shows the large number of *Vibrio cholerae* that can be present in a person with cholera.

lower intestine (colon) of all humans and many other large animals. Although the urinary tract is normally sterile (meaning there are no viable microbes present), it sometimes becomes infected, most commonly by variant strains of *E. coli* that form a particular type of pili, one that binds to the surface of cells lining the urinary tract. Similarly, *E. coli* strains that have a different type of pili cause some cases of diarrhea. Their pili bind to the surface of cells lining the ileum (upper intestine), where *E. coli* is usually not found.

In most of the mechanisms of dispersal that we have discussed, the microbe has no control over the distance and direction of movement, and the chance of a particular spore reaching a favorable habitat is very small. These microbes make up for this weakness by producing vast numbers of spores or similar propagating forms. Other microbes employ what might be considered a smarter strategy; as we shall see in the next chapter, they obtain information about the distribution of resources in their vicinity and exploit this information to guide locomotion.

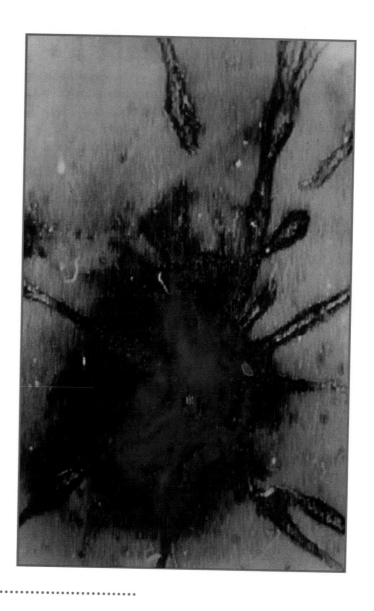

This false color image, taken under the microscope, shows cells of the cellular slime mold Dictyostelium (blue) streaming in columns in toward a central aggregate (also blue), to form a slug. To find their way to the center, the cells move against a spreading wave of increased cyclic AMP concentration, generated by the cells themselves

Navigating Through a Chemical Sea

As adept as many microbes are at locomotion, by itself the ability to change location has limited value. It becomes much more useful if the organism is also able to aim its motion in a favorable direction, toward a source of food or light. Without eyes or ears, a microbe still needs some way to sense the presence of food at a distance. It turns out that, for the smallest of microbes, with the simplest of sensory systems, the only way that an organism can find a source of food is by moving in random directions, and yet it is able to do so in a way that will inevitably lead to its goal.

Sensing Environmental Change

Probably the first sensory stimuli to be exploited during the evolution of life were chemicals that were needed as nutrients to feed cell metabolism. Primitive organisms developed mechanisms that allowed them to move toward higher concentrations of these chemicals, a behavior often referred to as "chemotaxis." Later, mechanisms evolved to respond to chemicals that were dangerous or associated with favorable or unfavorable conditions. Some mobile organisms became capable of navigating along a gradient of temperature—or light intensity—or by following a beam of light. Organisms that lacked sensory systems did not compete well with those that had them, and it appears that all organisms in our contemporary world have sensory abilities of one kind or another.

The major questions scientists ask about specific organisms fall into three groups—which we can categorize as "what," "how," and "why" questions: Precisely what stimuli or patterns of stimuli does an organism respond to? What is the response? How are the signals detected? How is the information processed? How does the organism decide what to do?

Why does the organism respond to these stimuli; that is, how do the stimuli relate to resources in the environment? Why does the organism respond in the way it does; how do the responses relate to the environment and how do they benefit the organism's ancestors?

Surprisingly, most progress has been made in answering the "how" questions. We now have a good basic understanding of the molecular and physiological mechanisms of most sensory systems. This impressive progress was largely possible because there are only a few basic mechanisms, which are used by many kinds of organisms for many different purposes.

It has proved much more difficult to determine what stimuli are used by organisms and to understand what information they obtain. Organisms are highly variable in these regards, and each species must be studied in detail. For example, the photosynthetic bacterium *Chromatium* is attracted to the electron donor, hydrogen sulfide (H_2S), which is a toxic repellent for most bacteria. Intestinal bacteria avoid benzoate, but the mud-dwelling bacterium *Rhodopseudomonas putida* can metabolize it and is attracted to it. Many microbes parasitic on plant roots are attracted to CO_2 (which roots produce), but most other microbes avoid it.

The detection of chemical stimuli usually depends on a chemical binding to special proteins called receptors and has much in common with mechanisms used to regulate chemical reactions within the cell. Given these internal regulatory mechanisms, it would be only a small step to develop receptors for detecting external stimuli, and this is the reason it is supposed that sensory systems for detecting chemical stimuli were the first to evolve. It appears that more organisms respond to chemical stimuli than to any other type.

For prokaryotes, zealous study of the mechanisms of chemoreception began in the 1960s with the work

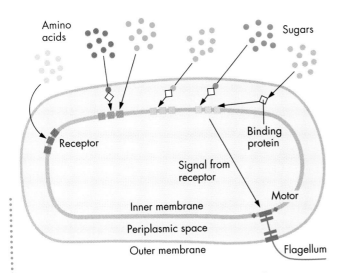

How bacteria sense chemical stimuli. Stimulus molecules (colored circles) bind to receptors (colored squares) embedded in the inner membrane, or they may bind to binding proteins (white squares) that in turn bind to the membrane receptors. In response, the receptor protein changes shape and thereby initiates a signal to the flagellar motor.

of Julius Adler and his associates at the University of Wisconsin on the intestinal bacterium *Escherichia coli*. Adler's prime strategy was to perform a "genetic dissection," a method that at the time was just being recognized as a powerful way of studying the molecular mechanisms of cells. In this strategy of research, genetic mutations are generated in normal organisms, and rare individuals with interesting characteristics are isolated and used to start true-breeding strains. The researcher then looks for molecular alterations in the mutant strains that may provide evidence about which molecular components are involved in generating the normal characteristic.

Adler's research group isolated mutant strains that did not respond well to normally attractive chemicals in a capillary tube. The researchers were then able to ascertain the number of genes involved in this behavior, the role of many of the genes, and the molecular components specified by the genes.

Both *E. coli* and *Salmonella typhimurium* (another extensively studied bacterium) are surrounded by two membranes enclosing between them the so-called periplasmic space. The outer membrane is relatively porous, and small molecules, including most chemical stimuli, readily pass through it. Many chemical stimuli bind to specific "binding" proteins located in the periplasmic space. These proteins then become activated signals and bind in turn to receptor proteins embedded in the inner membrane. Some chemicals bind directly to receptor proteins in the inner membrane, and certain sugars elicit a response after being transported directly into the cell.

Receptor Proteins in E. coli and S. typhimurium

Receptor protein	Stimuli	
	Attractants	**Repellents**
tsr	Serine	Acetate Benzoate
tar	Aspartate maltose-BP	Co^{2+} Ni^{2+}
trg	Galactose-BP Ribose-BP	—
tap	Dipeptides	—

-BP indicates that the receptor interacts with the sugar bound to a binding protein.

The best-studied receptor proteins belong to a group of proteins sharing a similar structure in part. One part of the protein lies outside the membrane, exposed to the periplasmic space, and varies in structure from one type of receptor to another. The bulk of the protein lies inside the membrane, exposed to the cytoplasm, and is very similar in all receptor types. A typical cell has hundreds of molecular copies of each receptor type, some more abundant than others. As much as one percent of the total membrane protein can be devoted to a particular receptor type.

When an appropriate chemical binds to an outside site on a receptor protein, it is thought to cause a change in the protein's shape. In response, a site on the protein inside the cell changes its interactions with other molecular components. This causes alterations in a chain of molecular interactions that ultimately alter the tendency of the molecular motors to turn in one direction versus the other. The details of this chain of interactions are still being unraveled through the efforts of many researchers.

Eukaryotes employ a distinctly different chain of reactions, although the receptor is still a membrane protein with a specific binding site exposed to the outside of the cell, and the binding of an appropriate chemical to this site still causes a change in the conformation of the protein. This change in turn causes the protein to interact differently with other proteins inside the cell, initiating a cascade of molecular changes within the cell. Only during the last decade have researchers found that most receptors for chemical stimuli outside a cell share a common molecular mechanism. This mechanism is used for signals involved in regulating development in all animals, as well as for external signals. The mechanism is altered in the development of cancer, and consequently a great deal of research effort has been directed toward understanding the mechanism in detail.

Choosing a Response

Whatever the nutrient sought after by a microbe, it is unlikely to be evenly distributed in space: in some locations it will be rarer, in others more common. Molecules released by, say, a nutritious colony of bacteria diffuse through space. The concentration of molecules, initially high near the colony, gradually diminishes with distance, creating a concentration gradient. Obviously it would be to an organism's advantage if it could move directly toward the location of highest

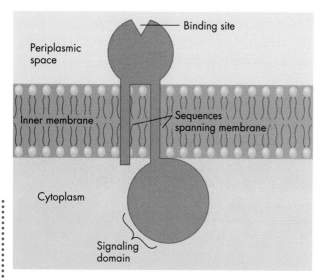

A receptor protein (green) is shown embedded in the inner membrane of a bacterial cell. The stimulus molecule attaches to the binding site in the periplasmic space, causing a change in the protein's shape that propagates through the membrane to the part of the protein that lies in the cytoplasm inside the cell. The change effected in the shape of the protein's signaling domain alters its interactions with other proteins, propagating the signal to the flagellar motor.

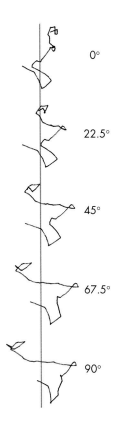

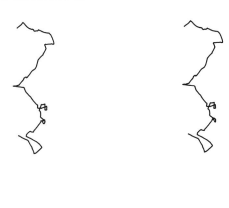

Left: Paul Frymier, using an apparatus built by Howard Berg, recorded this three-dimensional track of a swimming *E. coli* bacterium. Each dot corresponds to the cell's position, recorded every $\frac{1}{12}$ second for a total of 20 seconds. The track is shown rotated around the vertical axis a total of 90° to reveal its three-dimensional shape. Top: The track of a swimming *E. coli* bacterium, recorded for 41 seconds, presented as a stereo pair. You can see the three-dimensional shape of the track with a stereo viewer. Some people can see it without this aid by staring between the images and converging their eyes either behind or in front of the page. Each eye looks at a different image and the two images are fused into one. When this works, you see three images and the center one appears three-dimensional.

concentration. Yet a microbe has no way of knowing from its sensing of the concentration at a single location, which it judges by the number of receptors that are stimulated, in which direction the source of the nutrient molecule lies. To determine the direction to the source, it must measure the intensity at different positions in the stimulus field. It can then make progress toward the source by moving away from a position of lower intensity in the direction of a position of higher intensity.

To follow a chemical gradient to its source, a microbe relies on one of two fundamental methods. It can sample two positions simultaneously by using multiple receptors that are separated in different parts of the organism. In this case, the organism directly measures the *spatial* gradient by comparing the intensity at the different positions in the stimulus field. Alternatively, it can sample two positions sequentially by moving from one place or orientation to another, and measuring the concentration at each position. In this case, the organism directly measures a *temporal* gradient and infers the spatial distribution from knowledge about how its receptors were moved. Our visual system obtains detailed information about light distribution by a simultaneous sampling performed by the millions of photoreceptor cells in our eyes. In contrast, a dog follows a scent trail by sampling the air at many different times as it criss-crosses the trail.

To detect shallow gradients by means of a simultaneous comparison requires widely spaced receptors. Large body size is advantageous, sometimes achieved through the placement of receptors on antennae. Most bacteria are too small for simultaneous comparison to be effective and must make use of sequential sampling. Yet sequential sampling requires a coherent pattern of movement, such as locomotion in a straight line, whereas free-swimming bacteria are continually subjected to Brownian motion and caused to change direction in random ways that the cell can neither predict nor detect. Consequently, they can maintain coordinated movements for only a few seconds; in fact, free-swimming bacteria usually engage in behavior that causes their direction of locomotion to change every few seconds. Their path can then be described as a random walk, in which steps are taken in randomly chosen directions. Nevertheless, bacteria can work their way along a gradient by taking longer steps in one direction and shorter steps in the other. As a result, the random walk is biased in the direction of the stimulus field, a method of navigation termed "indirect guiding." The basic strategy can be described as "if conditions are improving, keep going, otherwise try a new direction."

Biased random walks along chemical gradients have been well analyzed in the intestinal bacteria *Escherichia coli* and *Salmonella typhimurium*. Individual cells of these bacteria swim rapidly when the motors of their multiple flagella turn in one direction. About once a second, however, the motors reverse direction, causing the flagella to change shape and separate from one another so that each flagellum pushes the cell in a different direction. The consequence is that the cell tumbles chaotically. After about 0.1 second, the motors resume rotating in the normal direction, and the cell swims off on a new course.

A bacterial cell makes progress along a stimulus gradient by modulating the timing of the motor reversals and the subsequent tumbling events. Careful experiments have measured the responses of individual cells that are exposed to pulses of chemical stimuli released from a nearby micropipette: the bacteria apparently average measurements of stimulus concentration taken over a period of about one second and compare this average to an internal signal representing the average concentration they have experienced over the previous 3 seconds. If the present average is more favorable, the tendency for the motor to reverse is suppressed, and the cell continues swimming in the same direction a little longer.

This signal-processing method makes good sense considering the restraints on the bacteria. If the memory extended back further than a few seconds, Brownian motion would have changed the direction in which the cell was swimming, and the cell would then be comparing the present concentration to concentrations at positions that have an unknown relationship with the present position. Because the comparison is confined to a few seconds, it is likely that the cell determined the remembered concentration at a position behind its direction of locomotion at the time of comparison.

If the period over which the cell averaged present concentration were shortened to be less than one second, then its measurement of concentration would be less accurate because fewer molecules would have bound to the receptors. It is always possible too measure concentration more accurately by taking more time. On the other hand, if the cell takes too much time, Brownian motion will cause it to shift its position before the averaging is completed. Thus, the one-second averaging time seems to be a good compromise.

Sensory Adaptation

When physiologists study sensory systems in animals, they usually find their subjects do not perceive a brightly lit scene to be much brighter than a moderately lit one; nor is the response of the system to sound proportional to stimulus intensity. Rather, eyes or ears adapt to a constant intensity and their response steadily wanes, but a quick change in intensity evokes a strong response. In fact, if optical tricks are used to fix an image of the external world on your retina, your perception of it will fade in a few seconds as the receptor cells adapt; normal vision depends on constant eye movements that sweep images across the retina. On a slower time scale, you become aware of sensory adaptation while entering or leaving a move theater, as your eyes adapt to darkness or light. We, and other organisms, sense changes in intensity rather than the actual level of intensity. We do so because changes are usually more important. For example, we care little about the intensity of sunlight but learn about objects by the manner in which their shape and material cause subtle changes in the intensity of reflected light.

Howard Berg and his associates have studied sensory adaptation in *E. coli* cells, using the same technique that they used to study the bacterium's motor: they cause the flagella to stick to a glass slide, so that the entire cell will rotate, then record the rotations. In this case, they recorded the rotations of a tethered *E. coli* cell in response to a 0.1-second pulse of chemical discharged by a micropipette within 4 μm of the cell. Counterclockwise rotation would propel a free cell forward, while clockwise rotation would cause the cell to tumble. Because *E. coli* cells change their direction of rotation spontaneously without any stimulation, Berg had to average the response over many repetitions before he could see a basic pattern. He de-

scribed the baseline behavior before stimulation as about a 64 percent probability that a cell is rotating counterclockwise—or moving forward—at any given time. After the attractant was released, this probability increased to a peak near 90 percent about 0.4 second after stimulation, declined steadily to a minimum value near 50 percent about 1.5 seconds after stimulation, and returned to the baseline value of 64 percent about 4 seconds after stimulation. Most cells rotating clockwise at the time of stimulation changed direction within 0.2 second. The tendencies for the probability of counterclockwise rotation to increase and decrease balance so that there is no response to a steady

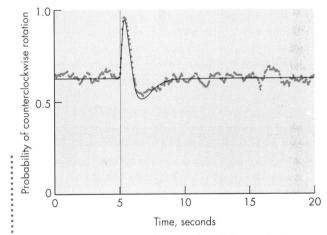

The response of *E. coli* to a pulse of chemical release at 5.06 seconds (gray line) on this scale, averaged over 378 experiments using 17 different cells. The pulse of attractant caused the probability of counterclockwise rotation to increase, which would lead the bacteria to swim on in the same direction, if they were not stuck to a slide. After the pulse dissipated, there was a decrease in this probability, or an increase in tumbling to change direction. By four seconds after stimulation, the bacteria's behavior had reverted to what it had been before stimulation.

Top: Multiple exposures taken 0.1 second apart, for a total of one second, reveal the behavior of the nematode *Caenorhabditis elegans* when it is held by the tail with a suction pipette. On the left, the nematode shows the normal behavior of forward motion: bending waves start at the head and spread to the tail, making the tethered nematode wave back and forth through a low amplitude. In the center, the nematode bends around to one side, a movement that would cause a freely moving nematode to turn around. On the right, bending waves starting at the tail and moving forward would cause a freely moving nematode to move backward, but this tethered nematode instead waves back and forth through a wide amplitude. Bottom right: To demonstrate adaptation, water was continually pumped past a tethered nematode; salt was absent from the water for 10 minutes and then present for 10 minutes in a repeating cycle. Nematodes are attracted to salt: its disappearance stimulates the nematode to initiate a reversal bout, while its appearance suppresses such bouts. Note that the response adapts; after a few minutes in a new salt concentration, the nematode's probability of initiating a reversal bout is nearly the same no matter what that concentration.

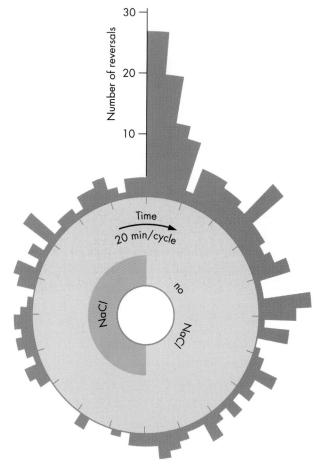

concentration. This is a form of sensory adaptation. Exposed to a pulse of repellent or to the removal of the attractant, a cell simply gave the inverse of the response to a pulse of attractant, with the probability of a counterclockwise rotation initially falling to about 15 percent and then rising to 70 percent.

In a similar fashion, I have studied nematodes held by the tail with a suction pipette. Surprisingly, although nematodes are much larger than bacteria, they also navigate by biased random walks, although they have other options, for reasons we will see later. In the experiment, chemical stimuli were added to water continuously pumped past the nematode, and the organism's behavior was recorded by photography or by projecting its shadow on an array of photodetectors connected to a polygraph. This technique permitted

the experimenter to control the pattern of chemical stimulation and record behavior with a one-second time resolution.

Subjecting the nematode to repeated cycles of exposure to the attractant NaCl revealed very clearly that when the attractant suddenly disappeared, the nematode was more likely to initiate a sequence of behaviors called a reversal bout, that in an untethered nematode would cause a change in direction and take it back into the salty environment which, in a normal situation, it might have inadvertently left. The nematode stops forward locomotion for a second or two, backs up for a few seconds, turns, and resumes forward locomotion. Such reversal bouts become more likely after a decrease in NaCl concentration, and less likely after an increase. Both responses adapted, and

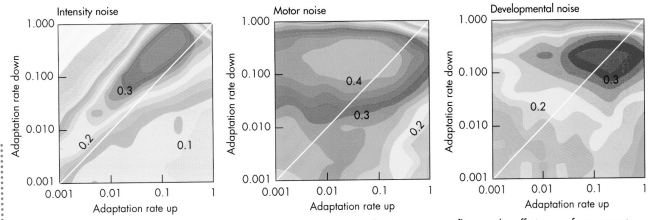

Computer models have been used to analyze how adaptation rates influence the efficiency of movement along a gradient, yielding charts like these that indicate the efficiency of movement using colors and associated numbers. The model considered three different types of "noise," or errors made by the organism, each having a different effect on the pattern of how adaptation rates influence efficiency. An organism may make systematic errors in measuring stimulus intensity (intensity noise) or random errors in attempting to travel a straight course (motor noise), or errors in the development of an organism may give it a tendency to turn in a particular direction (developmental noise). For intensity noise and motor noise, performance is optimum (dark blue) when the organism adapts faster to decreases in attractant than to increases—the pattern observed in nematodes.

the probability changed to the baseline level within a few minutes.

This response is very similar to that of the bacterium *E. coli* to a step-change in concentration. The main difference is that the nematode adapts over a period of a few minutes, while *E. coli* adapts in a few seconds. The advantage to the nematode is that it can measure concentration over longer time periods and improve its sensitivity. The nematode can afford to lengthen the time during which it measures the stimulus because it is held to the substrate and can maintain its orientation indefinitely, while the free-swimming bacteria are randomly reoriented by Brownian motion every few seconds.

Another interesting feature of the nematode response is that the nematodes adapted more rapidly to a dip in the level of attractant than to an increase. At first, I thought that this difference served no function and was probably an artifact of the adaptation mechanism. However, the same pattern of adaptation was seen in two very different nematode species, with different chemical stimuli, and even with temperature as a stimulus. When the same pattern was found in some experiments with bacteria, I formed the hypothesis that this difference in adaptation rates leads to more efficient movement along a gradient. A test of the hypothesis by computer modeling confirmed that this pattern of adaptation rates provides more efficient movement along a gradient than if nematodes adapted at equal rates to both kinds of stimulus change. Thus even the details of sensory information processing are fine-tuned to the needs of the organism.

A Cyanobacterium's Strategy for Finding Light

Larger microbes not knocked about by Brownian motion should be able to turn directly toward the stimulus source. It had been assumed that any organism large enough to be capable of simultaneous sampling would make such a "directed response." However, a recent observation suggests that the filamentous cyanobacterium *Phormidium* is an exception.

Cyanobacteria, also called blue-green algae, were one of the planet's dominant groups of organisms for eons in the past; even today there are thousands of species. These organisms resemble algae only in their ability to capture the energy of sun for use in photosynthesis. In their simple prokaryotic cell structure, they are far more like bacteria. Although each cell is fully viable on its own, in *Phormidium* and other filamentous cyanobacteria multiple cells form chains that become enclosed in delicate, tubular sheaths.

Filaments of *Phormidium* move by gliding. Like most photosynthetic organisms, they have developed sensory mechanisms to help guide them toward the optimal level of light available in their environment. *Phormidium* compares light intensities at the ends of a filament, which are separated by a substantial distance, yet it does not make directed turns. Rather, when the illumination is dim and the organism seeks light, it practices the following strategy: when its leading end is illuminated more strongly than its trailing end, it suppresses the frequency with which it changes direction; however, should the trailing end be the more strongly illuminated, it initiates extra changes in direction. This strategy ensures that the organism stays longer on a path taking it toward light and that it quickly leaves a path taking it in the wrong direction. Random variations in path direction (motor noise) probably keep *Phormidium* from becoming trapped oriented perpendicular to the light direction. This surprisingly sensitive prokaryote can respond even to moonlight.

Light, whether it is being made use of in photosynthesis or simply detected, must first be absorbed by a pigment molecule. Photosynthetic organisms are usually so full of pigment molecules (such as chlorophyll) that they absorb most of the light passing through them and appear colored. Researchers have wondered whether the same pigment is used by these organisms for both detecting light and capturing its energy for use in photosynthesis. Careful studies have been made of the colors of light that an organism responds to, known as its action spectrum. These studies usually reveal that the organism's movement toward light varies with color differently from the way that the absorption of light by its photosynthetic pigment varies with color. This is strong evidence that, for most species at least, the sensory pigment is different from the photosynthetic pigment. An advantage of this arrangement is that a distinct receptor pigment can be specialized to provide a more rapid response than the photosynthetic pigments, which are specialized for high efficiency.

Simultaneous Sampling

Many larger organisms are able to simultaneously compare the signals they receive from receptors located at different positions on their surface or to detect light coming from different directions. Using the information obtained about the orientation of the

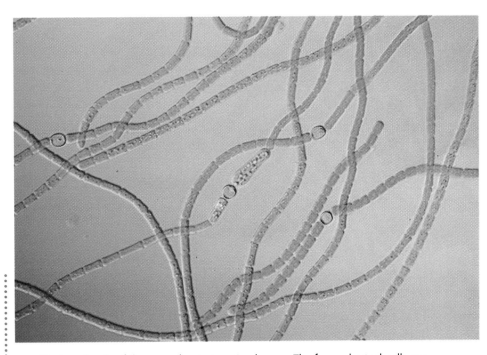

Chains of cells of the cyanobacterium *Anabaena*. The four spherical cells are heterocysts, cells that lack the apparatus for photosynthesis and are instead specialized for converting nitrogen taken from air to a usable form. The two granular and elongated cells flanking one of the heterocysts are cells specialized for surviving unfavorable conditions.

stimulus field relative to itself, an organism can make a direct turn toward a more favorable direction.

Anabaena is a common type of filamentous cyanobacterium which, as it glides in dim light, turns its leading end toward a light source, but in bright light turns away from such a source. Most photosynthetic organisms, in fact, avoid light that is too bright. Although exposure to light offers these organisms obvious benefits, light at high intensities can initiate damaging chemical reactions in the cell. Recent experiments by the German scientist Wilhelm Nultsch suggest that *Anabaena* has a mechanism to detect the presence of reactive forms of oxygen in the cell; the organism switches to light avoidance when the concentration reaches a certain level.

Experimenters who have shined small spots of light on *Anabaena* find that all the cells in a filament seem capable of responding. We assume that the organism relies on a simultaneous spatial comparison because the slow, smooth motion of gliding seems to preclude any mechanism of temporal comparison. The simplest hypothesis is that a difference in light intensity across the cell stimulates varying degrees of activity in the locomotor apparatus; with one part of each

A scanning electron micrograph shows various stages in the formation of the fruiting body of the cellular slime mold *Dictyostelium discoideum*. A migrating slug is in the lower left. When a slug stops migrating, it forms a clump that gradually elongates to finally form a thin stalk topped by a spherical bundle of spores, as seen on the right.

cell made to move faster than the other, the filament is forced to turn.

This behavior of *Anabaena* is an exception to the general rule that bacteria are too small to detect gradients. Because gliding bacteria are attached to a substrate they do not lose their orientation as a result of Brownian motion. Thus they can measure intensities over longer time periods and make more accurate determinations of light intensity—even across the width of a single cell.

Amoeboid cells of the cellular slime mold *Dictyostelium* nourish themselves on bacteria they have located by moving up gradients of folic acid, a chemical that is released by bacteria of many kinds. The amoeboid cells make simultaneous comparisons of the degree to which receptors scattered over the cell surface are stimulated; the most strongly stimulated areas are most likely to develop pseudopod extensions, and consequently the cell creeps toward the source of folic acid.

Cellular slime molds such as *Dictyostelium* grow as individual amoeboid cells until conditions for growth deteriorate. Then the cells aggregate to form a stalked fruiting body that bears spores at its tip. In the species *Dictyostelium discoideum* (isolated by Kenneth B. Raper in 1933 from decaying leaves in a hardwood forest in the North Carolina mountains) the aggregated cells migrate as a "pseudoplasmodium," or "slug," before forming the fruiting body. This slime mold has become a popular subject of study in the laboratory because it may represent an early stage in development of multicellular organisms.

Dictyostelium has taught us a great deal about the molecular components involved in chemoreception and motor responses, although we still do not know how all these components work together. An attractive hypothesis is that receptor binding generates two signals inside the cell. One is a long-range signal that spreads evenly throughout the cell, providing information about the average degree of receptor stimulation. The other signal is short-range and provides information about the degree of receptor stimulation in nearby parts of the cell. At any location in the cell, the ratio of these two signals would indicate whether that part of the cell is at the high or low end of the stimulus gradient, and this ratio would influence pseudopod formation or retraction.

Although *Dictyostelium* is not photosynthetic, the amoeboid cells are attracted to light of low intensity. When parts of the cell are illuminated to different degrees, the organism will move in the direction of the more strongly illuminated side, or in the opposite direction if the illumination is too strong.

When amoeboid cells of *Dictyostelium* aggregate to form "slugs," the mass of cells is sufficiently large to act as a lens that focuses the light impinging on a slug from one side to a higher intensity on the far side. The slug turns away from the side where the intensity is higher and toward the light source. This explanation is supported by experiments demonstrating that, if light is sufficiently absorbed in traversing the slug (because of added dye or use of ultraviolet light), the slug turns away from the light source, instead of toward it.

Sequential Sampling

Some organisms are able to make direct turns to a more favorable direction using only a single receptor. The trick is to move the receptor from one position or orientation to another, and compare the degree of stimulation at each position. If the movement is made in a coherent fashion relative to the organism's body, the organism can tell how the stimulus is oriented with respect to its own position, and it can make a biased turn directly to a more favorable direction.

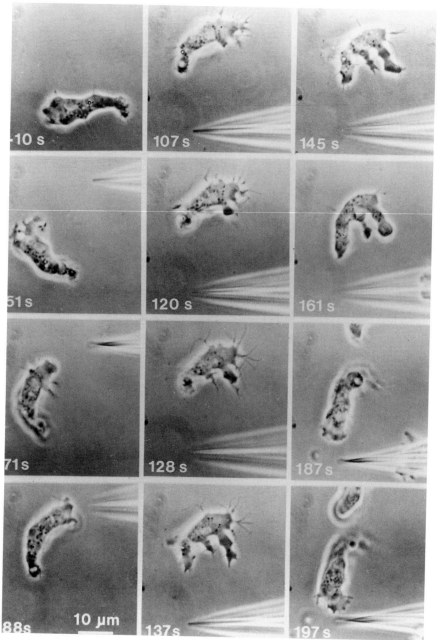

Testing the response of an amoeboid cell of *Dictyostelium* to the attractant cyclic AMP. Initially (−10 s) the cell was moving toward its broad end at the left. A micropipette containing a solution of cyclic AMP was brought in from the top right at time zero. (It appears as a set of V-shaped streaks because it is out of focus.) Within a minute (51 s) the cell had started extending pseudopods toward the pipette, and by the time 88 seconds had elapsed it had reached the pipette. At 100 seconds the pipette was moved to the lower edge of the frame. The cell now formed three main pseudopods extending toward the new position of the pipette (128 to 161 s). Within another minute, one of the pseudopods became dominant, and the cell moved to the pipette (197 s).

Unavailable to most bacteria, too pummeled by Brownian motion to make the coherent movements required, the mechanism is favored by most free-swimming eukaryotes (including single-celled ciliates and flagellates), which are sufficiently large that Brownian motion has relatively little impact on their locomotion. In swimming through open water, these organisms rotate around an axis and move along helical paths. This pattern of movement causes receptors for chemicals and temperature to be moved around the axis of locomotion (the axis of the helical path) and the receptors for light to be pointed successively in all directions at some angle to the axis of locomotion. This behavior provides the coherent movements of receptors required for sequential sampling and direct turns.

Consider an organism swimming northward through an east–west chemical gradient more favorable to the east and less favorable to the west. As it swims, its receptors will move a short distance back and forth along the east–west axis as well as along the up–down axis. In particular, its receptors will be most stimulated when it is to the east of the axis of locomotion, moderately stimulated when above or below the axis, and least stimulated when west of the axis. If the organism can compare intensities at different times during its rotation, each time making a small turn in the most favorable direction, it will turn gradually to the east until eventually it swims with the axis of locomotion pointing straight in that direction.

Nematodes, too, probably make use of this process at times, even though they also navigate like bacteria by means of biased random walks. Although nematodes possess a pair of chemosensory organs at their anterior end, one on each side of the body, the receptors cannot sample different positions in a stimulus gradient because nematodes lie on their sides, placing both sense organs at the same position in the

stimulus gradient. Hence the pair of receptors are assumed to function as a single unit.

As a nematode moves forward by undulating its body, its anterior end is moved back and forth across the substrate. Consequently, it can sample a gradient at positions on both sides of the general path of locomotion. It would be a simple matter for the nematode to bend its anterior end farther to the side should it sense more favorable stimulation. Done repeatedly,

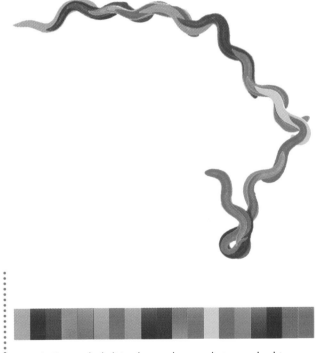

A *Caenorhabditis elegans* larva, photographed in rapid succession while crawling across an agar surface, makes a gradual turn, a right-angle turn, and finally a reversal of direction. With the help of a computer, I combined the images into one image using hue to distinguish the different images. The scale at the bottom shows the order in which hues were used from left to right. The images were taken 2 seconds apart during a 40-second period.

the sampling and bending would cause the nematode to gradually change its direction of locomotion until it was moving directly up or down the gradient. Some evidence suggests that nematodes do in fact make use of this process.

Why should nematodes employ two different means of navigation, at times relying on biased random walks and at other times on directed turns? A likely explanation is that, although directed turns lead the organism more efficiently to a source of nutrients, a stronger stimulus is required. There is only a limited interval of time during which the sense organs are held to one side and concentration can be measured, and yet the nematode must be sensitive enough to detect the small differences in concentration from one position to the next. This consideration, as well as some evidence, suggests that nematodes probably use the sensitive but inefficient mechanism of biased random walks when far from a stimulus source and the less sensitive but more efficient mechanism of directed turns when sufficiently close.

The most accurate form of steering known in microbes is one that guides many flagellated algae, which can orient to a beam of light and swim toward or away from the source in nearly a straight line. The ability is particularly important for algae, such as *Chlamydomonas,* that live in soil and migrate toward and away from the surface to stay within light of a favorable intensity.

Nearly all the wide variety of flagellated species exploit the fact that a cell moving through a three-dimensional medium will invariably rotate so that it travels in a helical path. These algae, which rotate at about one revolution per second, have developed structures that make the receptor most sensitive to light coming from a certain direction. As the cell swims and rotates, the directional receptor scans the environment much as some radar antennas do. If most light is coming from an angle to the axis of rotation, the cell will obtain a light signal that varies with the rotation. This changing signal guides these cells accurately toward or away from a source of light.

These directional receptors have only been appreciated in the last decade or two. The most obvious way to make a photoreceptor sensitive to direction,

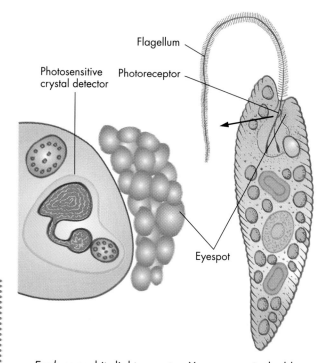

Euglena and its light receptor. You can see in the blow-up at left the photosensitive crystal light detector, which has a connection to the base of the flagellum. Changes in stimulation of the receptor somehow cause changes in the flagellum's pattern of beating. Adjacent to these structures is the so-called eyespot, which absorbs enough light to be visible in the light microscope; its purpose is simply to prevent light from one side from reaching the detector.

and a technique used by some algae, is to place the receptor near a light-absorbing structure that will absorb light coming from some directions before it reaches the receptor. Another mechanism that can be employed is dichroism—the differential absorption of light polarized in different directions. A given pigment molecule absorbs light most strongly when the light is polarized in a certain direction—that is, when the electric field of the light oscillates in a particular direction. If a light receptor is made up of pigment molecules all oriented in the same direction, then the receptor will respond more strongly to light polarized in certain directions. Even if the incoming light is not polarized—and most natural light is not—the receptor array will be more sensitive to light from certain directions.

Euglena and several other algae make use of a simple absorbing screen, called the eyespot, and a dichroic crystal receptor. Light coming from one side of the cell cannot reach the receptor because it is absorbed by the screening pigment. As the cell swims forward, it rotates around its long axis, and the side sensitive to light rotates also, scanning the environment. If the cell is not swimming directly toward or away from the light source, stimulation of the receptor varies during rotation, and the beating of the flagellum is altered in such a way as to cause the cell to turn in the direction that makes stimulation more uniform. Specifically, *Euglena* turns toward the eyespot and away from its flagellum shortly after the receptor enters the shadow of the eyespot. This response brings its locomotor path into alignment with light direction.

As effective as they are, these mechanisms of making receptors directional have one shortcoming in common: the receptors are not sensitive enough to detect dim light. Light coming from the correct direction makes only one pass through the receptor, and

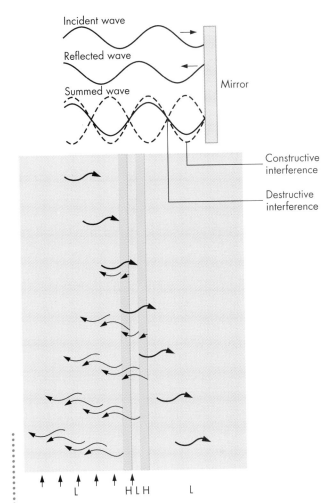

An incident light wave and its reflection interfere with one another (top) to produce a standing wave with alternating positions of constructive and destructive interference. A receptor pigment positioned where constructive interference occurs would experience greater light intensity. Some microbes, designed to take advantage of this, have reflectors formed from a stack of thin layers (bottom). Each boundary between layers of different refractive index reflects some of the light (black arrows) passing through it. The waves reflected from different layers interfere constructively and add up to a strong reflection.

only some of it will be absorbed—in dim light, not enough to be detectable. Many algae have developed a more sophisticated, and more sensitive, device that works by reflecting light. The device can be considered a kind of antenna because it makes use of diffraction and interference effects to obtain directional specificity, just as a television antenna does in capturing UHF or VHF waves.

The crucial element in these antennas is a thin-film reflector similar in design to many of the reflecting mirrors and nonreflective coatings now common in commercial optics. All such reflectors are built up of multiples of one basic unit, formed by two sheets of material placed together, one of high refractive index (light travels at a lower speed through it) and one of low refractive index (light travels at a higher speed through it). Simple as the unit is, it has an interesting, and highly useful, property: the interface between the two materials reflects a portion of any light incident on it. If a series of such interfaces are stacked together parallel to one another, light reflected from each layer will interfere with light reflected from other layers. If two light waves are out of phase, they interfere destructively and intensity is reduced; if two light waves are in phase, they interfere constructively and intensity is increased. If the stack of layers is such that each interface is one-quarter wavelength apart, then all the reflections will interfere constructively and a high-intensity reflection is produced. With a stack of many layers, almost all the light is reflected. The effects of reflective layers are commonly seen in nature. The colors in soap bubbles are created by the interference of light reflected from the two sides of the soap film. The bright iridescent colors of some insects are produced by light reflected from many such layers.

What makes a receptor with this device so effective in dim light is that the reflected light has a second

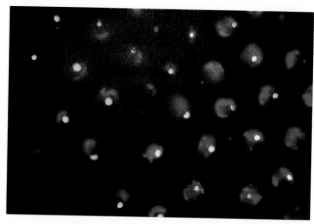

The reflection of light from the photoreceptors of *Volvox* cells. In this photograph, we see two dozen of the thousand cells of a *Volvox* colony, illuminated by blue light shining through the microscope lens. Some of the light is absorbed by chlorophyll in the cells and re-emitted as red fluorescence. Some of the light impinges on the light receptor at an angle at which it is strongly reflected, and we see a bright blue or white spot.

chance to pass through the receptor and be absorbed. Ideally the receptor pigment is placed where the incoming and reflected waves interfere constructively.

Iridescent colors have their characteristic changing hues because the wavelength of light that is most effectively reflected changes with orientation. Light traveling perpendicular to the stack of layers has the shortest path to travel between layers, and shorter wavelengths are reflected more strongly than in light traveling at an oblique angle. When combined with a receptor pigment that only absorbs shorter wavelengths, the structure becomes a highly directional receptor.

Looking through the light microscope with the appropriate illumination, you can see reflections from the eyespots of many species of green algae, such as *Chlamydomonas*. With the much greater resolution of electron microscopy, one can observe structures with the features of quarter-wave stacks in these eyespots. Careful analysis suggests that the quarter-wave stacks often work in conjunction with absorbing screens: light of wavelengths that are not efficiently reflected are absorbed so that they do not interfere with the detection of light direction. Some dinoflagellates have added a refinement to the system: they have a refractive lens that helps focus light on the quarter-wave stacks in the receptor. These peculiar eukaryotes, which are usually covered by stiff cellulose plates, are propelled by two flagella beating in grooves between plates, one groove oriented along the length of the cell, the other encircling it. Although dinoflagellates are often a major component of marine plankton, little is known about their behavior.

Volvox is a colonial green alga: each organism is formed of a few thousand flagellated cells arranged in a single layer to form the surface of a hollow sphere about 1 mm in diameter. Two flagella on the outside of each cell provide locomotion for the colony, and their motion causes the whole colony to rotate about an axis stretching from the anterior pole to the posterior pole. The steering mechanism of a colonial organism like *Volvox* faces an extra complication in its design: each cell acts as an individual unit, responding to light without regard to its neighbor's responses, and yet the combined action of the cells must move the organism in the required direction. The colony orients to a light beam by the independent action of

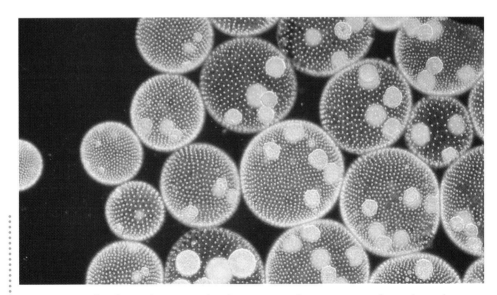

A group of *Volvox* colonies. Each colony is a nearly transparent sphere, the surface of which is formed of thousands of small cells, which are just barely visible in this photograph. In addition, each colony contains several large reproductive cells in its interior—these are the most noticeable structures.

the cells, which stop flagellar beating when light grows brighter and speed up when it dims. As a consequence, cells rotating from the shaded side to the illuminated side stop beating, and the continued activity of other cells rotates the colony so that its axis of rotation points toward the light source.

Do Microbes Form Images?

In most of the responses I have discussed, if an organism is faced with stimuli from two sources ahead, it will move along a middle path aiming toward neither source. If it moves sufficiently close to the midpoint, so that the sources are separated by a wide angle, it is then likely to move toward the stronger source. In contrast, many large animals can distinguish between two sources that are close together, choose one, and move directly toward it. The ability to distinguish between two sources of light has been considered to be the basic functional test for an ability to form images. Our question is: Can any microbes perform in this way?

Clearly, animals having vision are able to form images. A large array of receptor cells in the retina of the eye provides information on the intensity of light coming from many directions simultaneously. Since arrays of receptors are not known to exist in any microbe, one is tempted to reject this possibility. However, mammals can use their sense of hearing to

Drops of dew cover the surface of stalks of *Pilobolus*—the spore-throwing distance champion we encountered in Chapter 3. The black caps contain the spores and probably protect them from sunlight until they are eaten with the grass they stick to.

distinguish between sources lying in different directions, and in this case the receptors are confined to two ears that provide information on intensity and time of arrival at only two locations. Our sense of hearing can separate sources of sound by the brain's sophisticated processing of information from both ears. Microbes do not have brains, but they might have discovered some other trick to resolve multiple sources.

Fungi that propel their spores frequently face this type of problem. In many cases, the spore-bearing stalk is surrounded by obstacles, and it would be advantageous to aim the projectile at an opening in the vegetation. Openings can often be identified as sources of light, but in many cases, the fungus is presented with more than one opening. A fungus that averaged the sources together and aimed between the openings would not propel its spores very far. So, we may ask: If faced with two or more sources of light, do these fungi aim between them or do they pick out one?

In the spring of 1910, Hally Jolivette carried out the following experiment at the University of Wisconsin. She had a light-proof box constructed with two holes 1 cm in diameter, 9 cm apart, through which she directed light of equal intensity. She then placed a culture of *Pilobolus* 25 cm from the midpoint of the horizontal line between the holes. The two holes were separated by 20°, in the horizontal, when viewed from the location of the fungus.

Jolivette placed a glass plate in front of the holes, to which the discharged spore masses would stick, providing a convenient record of their position. In 36 experiments, she recorded the positions of 2504 discharges. The data clearly show that *Pilobolus* aims at one opening or the other and has an accuracy of about 1 cm, or 2°. Similar experiments demonstrated that

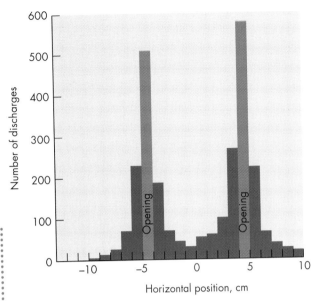

In 1910 Hally Jolivette measured the accuracy with which *Pilobolus* shoots its spore mass toward lighted openings of 1 cm diameter, placed 25 cm away. Most of the spores landed within one or two centimeters of one opening or the other. Clearly the fungi were aiming at only one of the openings and not between them.

with a separation of only 3 cm (7°), the spores still landed predominantly over the openings, rather than between them. This is a truly remarkable degree of discrimination.

But does this mean that *Pilobolus* can form images? No anatomical structures are known that could provide an imaging mechanism. On further consideration, it seems probable that our test for imaging is inadequate. What we really mean by imaging is the ability to distinguish between two or more sources simultaneously. In the experiment with *Pilobolus,* we have no evidence that when the fungus is aiming at one opening, it perceives the other. Rather, this experiment simply indicates that its aiming mechanism has a high degree of resolution. Put another way, its field of view may not be much larger than its accuracy. It can find the target, not by simultaneously sampling light intensity from many directions, but by gradually turning its narrow field of view in the direction of higher light intensity, something that could be accomplished using only three independent receptors. Eventually, the field of view becomes centered on a source that is bright compared to its surroundings but is not necessarily the brightest of all sources present. The achievement is still remarkable for a "lowly" fungus—even if it is not imaging.

Avoiding Obstacles

A microbe traveling in a normal manner toward a more favorable environment may become trapped behind a physical obstacle. In response, some microbes have developed specific behaviors to avoid obstacles. Microbes can easily steer around a physical obstacle if they know it's there—the trick is to detect the obstacle. In some cases, a microbe detects an obstacle only when it bumps into it, but other microbes can detect a physical obstacle at a distance.

When a swimming *Paramecium* runs into a physical obstacle, it backs up a few body lengths, pauses, and then goes forward again in a slightly different direction. If it encounters the obstacle again, it repeats the sequence of steps—if necessary, over and over until it has chosen a direction that does not lead to the obstacle. Jennings described this behavior as the "method of trial and error." Although it is a simple reflex behavior, it solves the problem of getting around a physical obstacle very effectively.

Paramecium has long been a favorite experimental organism, and one of its attractions is that its large

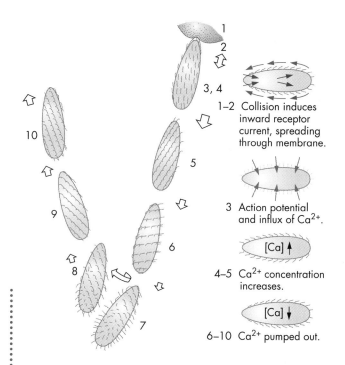

1–2 Collision induces inward receptor current, spreading through membrane.

3 Action potential and influx of Ca^{2+}.

4–5 Ca^{2+} concentration increases.

6–10 Ca^{2+} pumped out.

The behavior of *Paramecium* after colliding with an obstacle. The mechanical distortion of the cell opens membrane channels, setting off a depolarization of the membrane that spreads over the whole cell (1–2). Then channels permeable to the calcium ion are opened, and calcium flows into the cell, in which the element is normally present only in very low concentration (3). The calcium binds to proteins, causing a change in the cilia's pattern of beating, which in turn causes the cell to swim backward (4–7). After a few seconds, the channels close, the calcium is removed, and the cilia resume their normal beating pattern (8–10).

cells can be impaled with a microelectrode without suffering serious damage. Consequently, it is relatively easy to record changes in the electrical potential (voltage) across the cell membrane, and to measure net ion currents through the membrane. As a result, we now have a relatively detailed understanding of the mechanism of this behavioral reflex.

Bumping into an obstacle causes a channel in the *Paramecium* membrane to open and allows calcium and magnesium ions to enter the cell, setting off a change in membrane potential that spreads over the whole cell, similar to the action potentials of nerve and muscle cells; its wide reach solves the problem of coordinating the cilia in different parts of the cell. An influx of calcium ions into the cilia, stimulated by the change in membrane potential, changes the direction in which the cilia beat, and the *Paramecium* swims backward. After the internal concentration of calcium reaches a certain level, other processes restore the normal concentration.

Paramecium has a related response that helps it escape predators: touch the back end and it swims forward with a jump in speed. This response is also driven by the opening of membrane channels—in this case, channels permeable to potassium ions (K^+)—that set off a change in membrane potential. Thus, it appears that changes in a single parameter, membrane potential, coordinate all the various patterns of locomotion in this organism.

A *Paramecium*'s attempts to find its way around an obstacle are clumsy compared to the abilities of some other organisms. One of these is a fungus that has been famous since it came to the attention of one of the century's most renowned scientists, more than forty years ago.

The German physicist Max Delbrück, who contributed half a dozen papers on quantum mechanics in the 1920s and 1930s, was inspired in 1932 to set off in a new direction by a lecture he heard Niels Bohr deliver, suggesting that the study of biology might lead to new concepts in physics. After many fruitful years of research in genetics, during which he is credited with initiating the study of molecular genetics in general and bacterial genetics in particular, Delbrück turned to sensory physiology in his search for new

concepts in physics. In 1953, he chose the fungus *Phycomyces blakesleeanus* as the best model for studying sensory physiology.

Of all the organisms in the world, why did he choose *Phycomyces*? All microbes share the advantage of simplicity of structure compared to plants and animals, but in the 1950s techniques for observing the behavior of microbes were very limited. *Phycomyces* was attractive for study because it forms exceptionally tall, but simple, spore-bearing stalks.

In nature, *Phycomyces* is found growing on decaying organic matter. It is among the earliest fungi to appear on wet feces, although it is later replaced by slower-growing fungi that can degrade cellulose. Our best guess is that its spores are ingested by mammals along with food, pass through the digestive tract, and germinate in the feces. Where *Phycomyces* appears, it attracts attention by its exceptionally tall spore-bearing stalks, which, although only 0.1 mm thick, can be as much as 100 mm long. It is one of the few fungi that can produce simple stalks that are able to stand up in dry air, provided only that the mycelium is damp.

These spore-bearing stalks, called sporangiophores, are tubes with a single cylindrical cell wall. They elongate by adding material to the cell wall in a "growing zone" 0.1 to 3 mm below the spore mass. The cell wall grows in a twisting as well as an elongating fashion, causing the tip of the sporangiophore to rotate.

Presumably the function of the stalk is to get the 100,000 spores carried at its tip away from the decaying material and out into the open where they can be dispersed. The mechanisms of dispersal are unclear; *Phycomyces* does not release dry spores that can be carried by the wind. Nevertheless, biologists recognized as early as the nineteenth century that the sporangiophores grow relatively rapidly (at about 1 μm/s) and that their growth is guided by light and a variety of other stimuli. The tendency of an organism to grow toward or away from a stimulus is often called a "tropism." Intrigued by the phenomenon, biologists studied *Phycomyces* behavior extensively even in the nineteenth century.

Phycomcyes sporangiophores grow

- against gravity (negative gravitropism),
- toward light (positive phototropism),
- into wind (positive anemotropism), and
- away from physical obstacles (fugotropism).

All these responses can be understood as mechanisms to steer the growing sporangiophores away from obstacles and to get the spores out into the open. Delbrück was interested in the quantitative problem of how biological sensory systems manage to respond to sensory signals, like light, over enormous ranges of intensity, from very dim to very bright. Since he was a famous scientist, commanding substantial resources, the *Phycomyces* sporangiophore has become one of the most carefully characterized sensory systems in all of nonhuman biology.

Delbrück might have chosen differently if he had known the difficulties he would face, to this day unresolved. Although *Phycomyces'* response to gravity is well documented—in the weightless environment of an orbiting spacecraft, sporangiophores bend randomly and form irregular loops—Delbrück chose light as the subject of his studies of sensory physiology, because light is the most easily controlled and measured stimulus. Nevertheless, after decades of studying the response of *Phycomyces* to light—since 1970, with lasers and computer controls—scientists are not much closer to answering Delbrück's original questions. They have, however, come to appreciate the

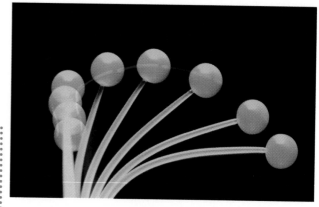

This long exposure photograph shows a *Phycomyces* sporangiophore responding to light. Every four minutes, David Dennison of Dartmouth College illuminated the sporangiophore with a flash of orange light, a color of light that *Phycomyces* does not detect. During the same period he had a constant blue stimulating light directed onto the sporangiophore from the sides (you can see its reflection off the head of the sporangiophore). Initially, stimulus lights of equal intensity came from both sides, and the sporangiophore grew straight up. After the third flash, Dennison turned off the stimulating light from the left and doubled the intensity of light from the right. Within a few minutes the sporangiophore began to grow toward the right, into the stimulating light.

fungus's impressive sensitivity to light. The dimmest light to which it responds (10^{-9} W/mm^2) is about a million times less intense than moonlight. *Phycomyces* can grow toward an opening to the sky even on a dark night, when the opening would not be visible to a human being.

Moreover, scientists have learned something about the physical mechanisms of bending toward light. When a sporangiophore is illuminated from one side, it bends toward the light until it reaches an inclination at which its tendency to bend toward the light is balanced by its tendency to bend away from gravity. The bending comes about by differential growth: the side away from the light source grows faster, and the side facing the light source grows slower. This is the opposite of what we would expect judging from the sporangiophore's faster rate of growth in balanced light. The explanation is that the sporangiophre acts as a cyclindrical lens that concentrates light in a band along the far side of the sporangiophore, stimulating the growth of the cell wall in this part of the cell.

Many detailed experiments indicate that the sensory-response system is quite complex. There are probably two sensory systems, one specialized for low light intensities and one for high. There appear to be different adaptation mechanisms for increases and decreases in light intensity, and for the sensory and response sides of the system. Analysis of the system is complicated by the fact that there is no good spatial frame of reference, since the sporangiophore is constantly elongating and twisting, and some components of the system move relative to others during the response. Thus, what was originally assumed to be a simple model system now appears to be a very complex system that may rely on specific subtle effects for its high level of performance. The search initiated by Delbrück for the secret of the ability to respond to a wide range of intensities may never come to a satisfactory conclusion, nor is there any hint of unusual physics being involved.

An even more fascinating behavior of *Phycomcyes* is its avoidance response. If a solid object is placed within a few millimeters of the growing zone, the sporangiophore bends away from it. After a delay of 2 to 3 minutes, bending increases for half an hour, by which time the sporangiophore may be growing 45° from the vertical. Eventually, as the growing zone moves farther from the object, the sporangiophore

bends back toward the vertical. If the object is moved to remain close to the growing zone, the response continues without adaptation. Placing objects symmetrically on both sides of the sporangiophore causes it to increase its growth rate without bending. If the growing sporangiophore encounters a surface inclined to the vertical, it will avoid the surface and grow parallel to it 2 to 3 mm away. This response has obvious benefits in guiding the growing sporangiophore away

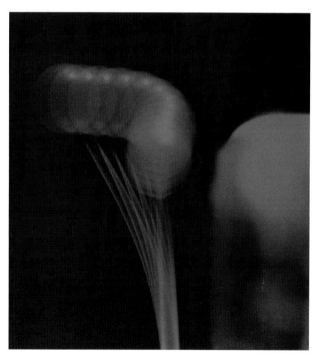

A multiple exposure photograph of *Phycomyces* avoiding an obstacle (an exposure was taken every two minutes). David Dennison illuminated the sporangiophore with red light, which is undetectable to the fungus. After five exposures (10 minutes) he moved an object up to the right side of the sporangiophore, but not close enough to touch it. In the next few minutes, the sporangiophore clearly bent away from the object.

from physical obstacles. The big question is: How does the sporangiophore sense the presence of the object without touching it?

This question has challenged biologists for several decades, and many talented students and professors have struggled with it without obtaining a clear answer. The type of material composing the object seems to be irrelevant—and every conceivable type of material has been tested, including glass, quartz, salts, plastics, Teflon, metals, magnets, crystals transparent to infrared radiation, activated charcoal, water, paraffin oil, and fluorocarbon oil. The object can be very small; a thread only 15 μm in diameter is avoided if placed sufficiently close to the growing zone. On the

other hand, a moving surface (such as a cylinder rotating around its axis) is often not avoided.

To explain these observations, biologists have proposed that the sporangiophore emits a volatile chemical that is carried away by diffusion and the movement of air currents. Obstacles exert their effects by stilling air currents; no longer dispersed by moving air, the chemical increases in concentration on the side toward the object. This hypothesis led to the prediction that placing a small enclosure around a sporangiophore, to reduce air currents, would cause a spurt of growth. And, indeed, when a glass enclosure a few inches in size is lowered over a sporangiophore, it responds, after an interval of 2 minutes, with a transient increase

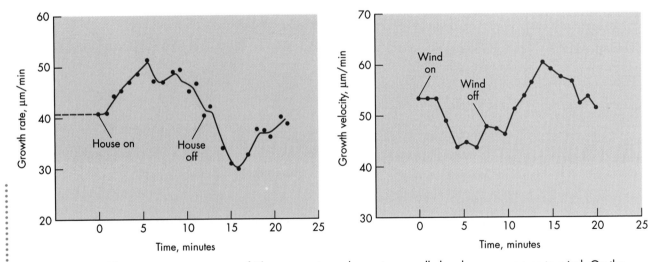

The contrasting responses of *Phycomyces* to enclosure in a small chamber or exposure to wind. On the left, a chamber measuring 5 × 5 × 7.5 cm (called a "house" by the researchers) was repeatedly lowered over the sporangiophore for 13 minutes and raised away from it for 10 minutes. Averaged for three cycles of this simulation, the data show that enclosure causes an increase in growth rate for several minutes, until the fungus adapts. Removing the enclosure causes the growth rate to slow, again until the fungus adapts. On the right, a similar experiment exposed *Phycomyces* to gentle (3 cm/s) horizontal wind, which blew for 7 minutes (too short a time for much bending to occur) and was off for 12 minutes. The figure shows data averaged over five cycles of simulation. The wind caused a decrease in growth rate, just as removing the enclosure did. These observations suggest that enclosures and barriers exert their effects primarily by stilling air currents.

in growth rate lasting about 10 minutes. The enclosure retains most of its effectiveness even if made of wire screen, as expected if the effect is due to the dampening of air currents.

Another prediction of the hypothesis is that experimentally induced air currents would cause the sporangiophore to slow its growth rate. Again, a transient response was observed, whether the air moved horizontally, upward, or downward. Horizontal currents also induce bending—*into* the wind. The threshold wind speed is only about 1 cm/s, which is so low that it is difficult to eliminate stimulating air currents even in the laboratory.

Howard Berg and two associates at the California Institute of Technology attempted to address this problem by building a chamber with a small temperature gradient so that cool, dense air would stay below warmer, lighter air. In this apparatus, the air currents moved at less than 10 μm/s, much lower than the speed at which molecules diffuse through air. Consequently, the residual air currents should not have had any effect on the distribution of molecules.

After overcoming technical problems in getting the sporangiophores to respond in the chamber, Berg and his associates found that the sporangiophores could avoid obstacles even in the absence of air currents. Most revealing, at a distance of 0.5 mm, a thin glass fiber was as effective at bending the sporangiophore as a large flat glass plate, although the latter provided an effective stimulus at greater distances (up to 4 mm). These observations—that the strength of the response to a large flat barrier does not change rapidly with distance and that a small fiber is nearly as effective—conflict with the hypothesis that the barriers simply reflect a chemical that the sporangiophores grow away from.

These investigators favor a model in which the sporangiophore releases inert molecules with a very special property: if the molecules adsorb to a barrier, they are converted to a growth stimulator with a limited lifetime and released from the barrier. Some of these molecules diffuse back to the sporangiophore and stimulate growth predominately on the side facing the barrier. One reservation is that no one has demonstrated that a wide variety of materials are avoided in the absence of air currents. Thus, it may well be that either air currents or specific chemical properties are sufficient to change the concentration close to the sporangiophore. Until the molecules involved are identified, the mechanism of this fascinating behavior will remain uncertain.

However, the mechanism of a similar behavior in a different organism has been worked out. Growing fruiting bodies of the slime mold *Dictyostelium* show a similar kind of avoidance behavior within 0.8 mm of a barrier. In this case, although they avoid most surfaces, they move toward an adsorptive barrier of charcoal, suggesting that the fruiting bodies respond to a chemical reflected from barriers formed by normal materials. There is evidence suggesting that ammonia, NH_3, is the chemical that *Dictyostelium* uses to probe the environment. It has been determined both that ammonia is released and that the fruiting bodies avoid ammonia.

The sophisticated behaviors described in this chapter can guide microbes toward a goal lying in any direction whether up, down, or to either side. Many microbes living in the soil or in pools of water, however, move about mostly along a vertical axis— toward or away from light at the surface above, for example—and must find the depth that best satisfies their needs. Under these circumstances, the pull of gravity, gradients of temperature, and even the earth's magnetic field, offer microbes new ways to guide their locomotion.

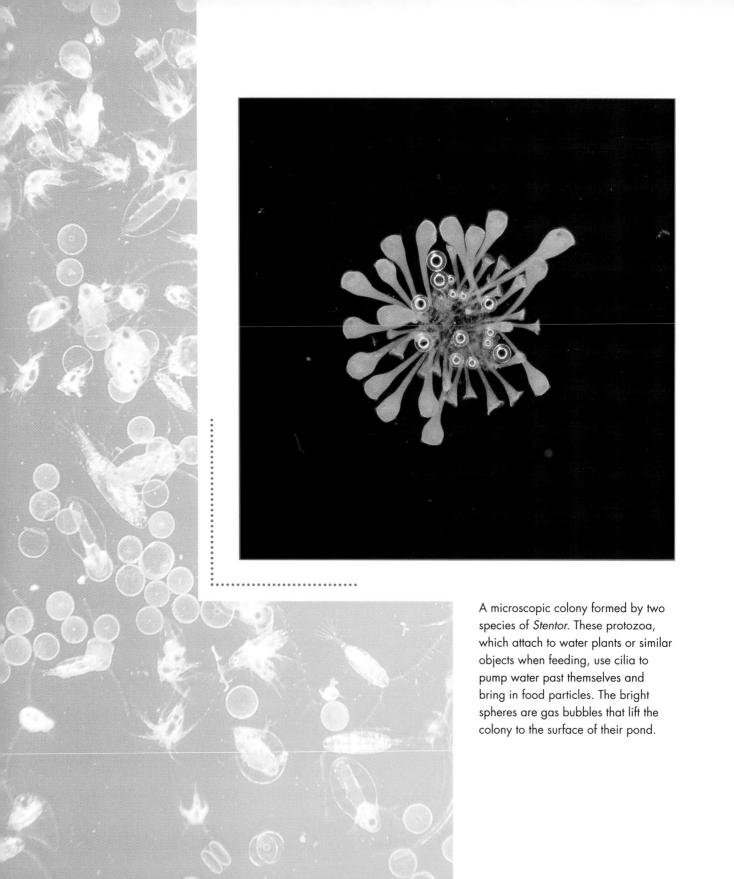

A microscopic colony formed by two species of *Stentor*. These protozoa, which attach to water plants or similar objects when feeding, use cilia to pump water past themselves and bring in food particles. The bright spheres are gas bubbles that lift the colony to the surface of their pond.

CHAPTER **5**

Vertical
Migration

Sunlight and oxygen are readily available at the surface of ponds, oceans, and soil, but diminish rapidly with downward movement. Consequently, many microbes migrate up and down to place themselves in the appropriate light intensity or oxygen concentration. For example, cyanobacteria that live in marine and fresh waters, or in mud, retreat from the surface during the brightest times of the day and come back up when the light is less intense. A few centimeters of water will screen out damaging ultraviolet light, while visible light may penetrate many meters. Only 3 mm of sand, 0.5 mm of microbial mat, or 0.2 mm of mud reduce visible light intensity a hundredfold; in these environments a microbe that travels less than a millimeter may find itself in significantly different conditions.

Oxygen enters the water from the atmosphere but is consumed by microbes. Where the water doesn't mix much vertically, because of thermal stratification or the proximity of solid surfaces (as in saturated soil or sediments), microbes often consume oxygen faster than it is transported from the atmosphere, and the environment becomes anaerobic—lacking in oxygen. The effects on organisms of all kinds are profound because oxygen is required by many organisms but, in large enough doses, is toxic to all. Most organisms are specialized to perform well either in atmospheric levels of oxygen, or in lower levels of oxygen, or even in the complete absence of oxygen, but few can grow in all these conditions.

Floating in the oceans are huge numbers of microbes that make up the plankton. Caught up in currents, these organisms are unable to move under their own power to new environments in horizontal directions; but their short vertical movements often

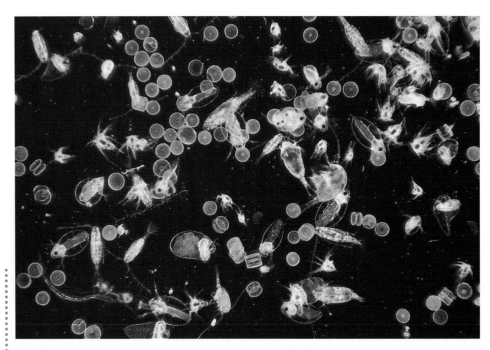

A typical sample of plankton, containing diatoms (the short cylinders) and various micro-crustaceans (the creatures with jointed appendages).

change the environment dramatically. Cyanobacteria, diatoms, and dinoflagellates, called phytoplankton, are photosynthetic members of this community. The non-photosynthetic members, or zooplankton, comprise single-celled protozoa as well as some multicellular, though still invisibly small, rotifers and micro-crustaceans. Moving up and down presents these organisms with new challenges: those that have flagella can swim up or down but need some way to keep their bodies correctly oriented. Others use methods of locomotion especially adapted for vertical movement. Some are simple, like a change of buoyancy, and others are complex, like the use of temperature gradients to navigate through the darkness of soil.

Changing Buoyancy

Photosynthetic microbes have a clear reason for being near the surface in the daytime, at least when it's not too bright, but many head back into the depths after sunset. Why bother migrating downward at night? An attractive hypothesis is nutrient retrieval.

Phytoplankton often face the dilemma that light for photosynthesis is most available at the surface, but all the cyanobacteria and algae growing at the surface deplete those waters of nutrients. Nutrients often remain available at greater depths, however. Since there is no light available at night for photosynthesis anyway, it might be productive to move down to these richer waters at night and obtain nutrients for use during the next day.

Planktonic organisms with flagella can swim to the appropriate depth, but those without these appendages need some other mechanism for maintaining an appropriate depth. Many planktonic microbes that lack swimming abilities solve this problem by regulating their buoyancy.

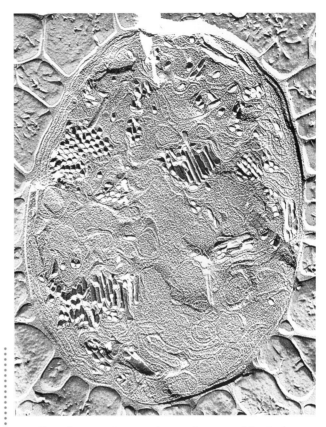

This electron micrograph reveals gas vesicles in the cyanobacterium *Anabaena flos aqua*. The vesicles are the clumps of parallel cylinders, some cut perpendicular to the axis so they appear circular and others cut parallel to the axis so they appear elongated.

Cells normally sink because most cell components are denser than water. Whereas water has a density of 1.00 g/cm³, nucleic acids (DNA and RNA) have a density of 1.7, carbohydrates a density of 1.5, and proteins a density of 1.4. A cell can be made buoyant by storing sufficient lipid or fat (0.90 to 0.94 g/cm³), but this is a very costly investment. A more efficient method of reducing cell density is to form gas vesicles.

Most planktonic cyanobacteria form gas-filled vesicles that allow them to float. In the still waters of a

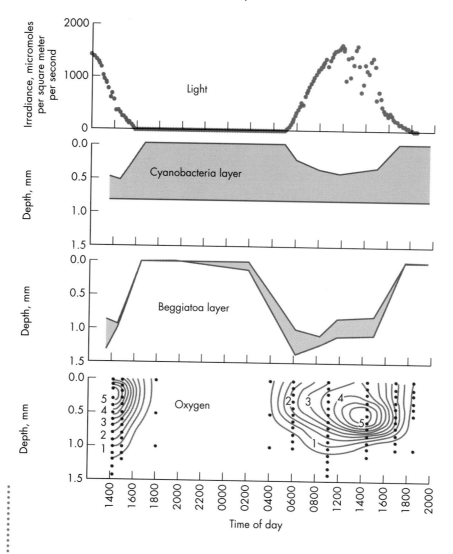

Often, microbes need migrate only very small distances, as demonstrated by a detailed study that investigated a microbial mat on the bottom of a salty pond in Baja California. At night cyanobacteria extended to the surface, but during the brightest part of the day the cells at the surface migrated about half a millimeter into the mat to avoid damaging light intensities. The sulfide-oxidizing bacteria *Beggiatoa* migrated about a millimeter to avoid high oxygen concentrations produced by the photosynthesizing bacteria. Photosynthesis drives the oxygen concentration above saturation (as indicated in the lowest graph, in which the digits give the number of times by which the oxygen concentration exceeds saturation). Note that the concentration is highest at a depth of half a millimeter (at 1400 hours), just at the top of the layer of cyanobacteria.

lake stabilized by thermal stratification, diatoms and nonflagellated green algae are forced to sink by their own body weight. In contrast, cyanobacteria often flourish under these conditions: by regulating their buoyancy, they can maintain themselves at an optimal depth or migrate between different depths.

The gas vesicles are remarkable structures, formed from a single type of protein—much like the shell of a virus. They are typically cylindrical with conical ends and are highly permeable to gases but impermeable to water. During assembly, they grow from a small empty vesicle that never contains water and, when fully formed, have a density of only 0.13 g/cm³. If a pressure much greater than normal is applied, the vesicles collapse to a tenfold higher density and cannot be reinflated.

There is some evidence suggesting that cyanobacteria can regulate their buoyancy—at least crudely. Vesicles are made continuously in sufficient quantity to make the cells float. A variety of vesicles that collapse at somewhat different pressures are present. When a cyanobacterium is transferred to high light intensities, photosynthesis speeds up and many small molecules accumulate in the cell, increasing its osmotic pressure. The pressure collapses some of the gas vesicles, causing the cell to sink.

Computer models have demonstrated that changes in gas vesicles generally cause the cyanobacteria to move toward levels of optimal light intensity. Although a small single cell can sink or rise only a few centimeters per day, a compact group can generate more buoyant force in proportion to its hydrodynamic drag, and colonies several hundred micrometers in diameter can move many meters up and down during a day. The actual pattern and depth of migration depends on the rates of metabolic activities and on colony size. Thus, populations can specifically adapt to conditions that are maintained for some time.

A study of one-meter-deep fish ponds in Israel has demonstrated that body weight alone can move microbes downward and upward. Cyanobacteria formed a scum on the surface of these ponds during the morning that disappeared later in the day. These bacteria increased the amount of carbohydrate they carried by threefold during their morning photosynthesis sessions at the surface. Made denser by the added carbohydrate, the microbes sank to the bottom where nitrogen was available. During the night, their carbohydrate was depleted—presumably because the bacteria used energy from the carbohydrate and nitrogen from the environment to synthesize proteins and nucleic acids, and excreted carbon dioxide waste. The loss of carbohydrate reduced their density, and they rose to the surface by morning.

The Complicated Migrations of Zooplankton

Whereas photosynthetic microbes are commonly found near the surface during the day and deeper at night, the pattern is the reverse for most zooplankton. Some migrating zooplankton are reacting to the migrations of their photosynthetic neighbors. Photosynthetic activity can produce supersaturated oxygen concentrations that are very toxic to some microbes. During the day these sensitive organisms migrate deeper to avoid oxygen toxicity, and at night they move to the surface to obtain normal oxygen concentrations. The larger zooplankton, however, seem to migrate primarily to avoid being spotted by fish looking for a meal.

Many features of aquatic organisms seem to be geared to reducing their visibility to fish, most of which hunt by sight. One way for an organism to become less visible is to be transparent. Many organisms living in open water are much more transparent than organisms in other habitats. This method of camouflage is not available to photosynthetic organisms, however, since they must be pigmented in order to capture light.

A simple way to avoid being seen is to avoid lighted areas. Where water is deep, it is always dark if you go deep enough. However, most food is generated at the surface, where sunlight is plentiful. Thus, organisms large enough to be hunted by sight face conflicting demands. Zooplankton respond by making daily vertical migrations, spending daylight hours at depths where the darkness protects them, but moving toward the surface, where more food is available, at night.

Maciej Gliwicz of the University of Warsaw measured the depth favored by the micro-crustacean *Cyclops abyssorum* at noon and at midnight in six deep lakes in the Tatra Mountains of Poland. The *Cyclops*

migrated less than a few meters in the two lakes that did not contain fish and about 10 meters in three lakes that had been artificially stocked with fish five to thirty years before. But they migrated about 20 meters in the lake that had been populated by fish for many centuries. The more time the *Cyclops* had to evolve defenses to fish predation, the greater the extent of their daily migrations.

Organisms in a long food chain may engage in complex vertical migrations, as those in the middle of the food chain seek to obtain food without becoming food. For example, rotifers might favor habitats containing fish that preyed on the rotifers' own predators, mostly micro-crustaceans, but not on the smaller rotifers. Meanwhile, the small fish that prey on the micro-crustaceans are also migrating in order to find food without being eaten by larger fish.

As if this were not complicated enough, phytoplankton, in addition to migrating up and down in

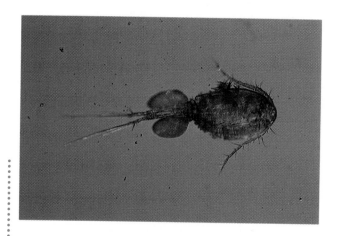

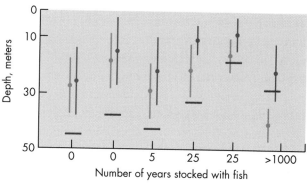

Left: The copepod *Cyclops* gets its name from its single eye, seen as the red spot at the tip of the head. The two green bags at the rear carry eggs. Right: In six deep lakes of the Tatra Mountains, *Cyclops* copepods were found to migrate longer daily distances in lakes that had contained fish for many years. Researchers took samples at noon (red) or midnight (blue) from various depths and recorded the mean depth (circles) and extent of depths (vertical lines) at which *Cyclops* were found. The *Cyclops* were usually found above the level marked by the horizontal bars, the depth at which sunlight is attenuated to one percent of its surface intensity and above which there would be more algae to feed on.

search of light and nutrients, may also migrate to avoid zooplankton predators. Most phytoplankton are too small to be visible to predators, and the small predators that eat them seem to be guided to phytoplankton by sensitivity to water currents. Thus the phytoplankton could be avoiding predators sensitive to currents, which in turn are avoiding predators guided by sight. In any case, daily vertical migrations are common among animals larger than about a millimeter. Some rotifers complete a vertical migration of a meter or so up and down daily.

Steering to Gravity

Organisms having to accomplish such complex migrations need to distinguish up from down. Some organisms use the direction of light as a cue to the upward direction, but what can be done in the dark? One possibility is that microbes, like many familiar animals and plants, maintain their proper orientation even in the dark by determining the direction of gravity.

For a long time there was no convincing evidence that any microbe could *sense* the direction of gravity, but some can swim preferentially upward or downward as a result of gravitational and viscous forces acting on the whole organism. If the center of gravity and the center of viscous resistance are offset from each other, the two forces will tend to rotate the organism until they are pulling in exactly opposite directions. For example, if the leading end is more dense than the trailing end, the cell will tend to rotate to a position in which it is heading downward. This is the same principle by which lightweight, high-drag feathers on the rear of an arrow help keep the arrow flying straight.

Most organisms are denser than water and tend to sink. If they have an asymmetrical shape (for instance, a high-drag flagellum sticking out of a spherical cell),

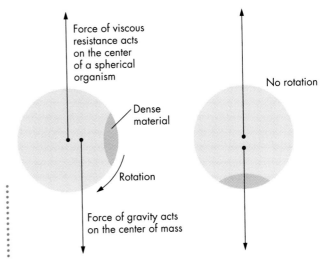

An uneven distribution of density can cause even a spherical organism to orient to gravity. As an object sinks, the drag exerted by the water is determined solely by the shape of its surface. In contrast, the force of gravity acts on the center of mass. Consequently, the two forces can act at different positions and exert a turning force (torque) on the organism that causes it to rotate into a position where the two forces pull in exactly opposite directions.

hydrodynamic forces acting on the sinking organism will tend to orient it with the compact end downward (or the flagellum trailing upward). This tendency will be superimposed on any swimming motion. Other factors being equal, a cell with a flagellum that pulls the cell forward will tend to swim upward, while a cell with a pushing flagellum will tend to swim downward. For example, cells of *Chlamydomonas reinhardtii* have been measured to swim at a mean speed of 60 μm/s, while they sink at 1 to 30 μm/s. Long uninterrupted swimming paths tend to curve upward at about 10°/s. To move to a new depth, a cell can sink downward or swim upward.

The mechanism discussed so far can work even if all parts of the organisms have the same density. In reality, different cell components have somewhat different densities, and even a symmetrically shaped organism would orient under gravity if its density were different at one end than at the other. The nematode known as the vinegar eel *(Turbatrix aceti)* is an example: it tends to swim upward because its tail end is denser than its front end.

Dinoflagellates often move upward during the evening and downward during the morning. They seem to descend faster than their sinking speed, which suggests that they are actively swimming downward, presumably oriented by the forces of gravity and viscosity acting together on the cell. Interestingly, the organisms begin their upward migrations before dawn, and in the laboratory they migrate even when kept in complete darkness. The cells seem to make use of an internal clock to anticipate changes in lighting. In addition to the clock, though, the cells also need some way to reorient themselves for the upward swim. A shift in the center of gravity, perhaps achieved by changing the amount of carbohydrate or fat stored at one end, could accomplish the reorientation.

In all of these mechanisms, gravity is supplying the force that rotates the organism. This is in contrast to the orientation of arthropods and vertebrates, to which gravity supplies only information, via a specialized sense organ; the organism uses its muscles to cause the rotation. Do any microbes have such a sense organ and use this kind of stimulus–response reflex to orient to gravity?

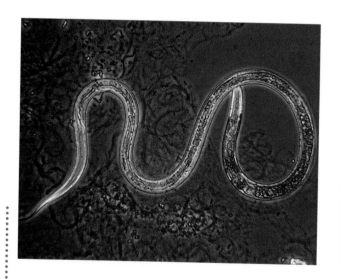
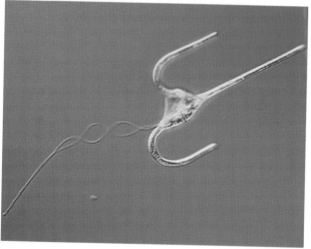

Left: The nematode scientists call *Turbatrix aceti* is called the vinegar eel by others because it lives in fermenting solutions where bacteria are producing acetic acid to form vinegar. In the past, it was commonly found in commercial vinegar. The nematode is unique in that its unusually high length-to-width ratio of 45 (compared to about 20 for most nematodes) allows it to swim more efficiently, and it swims well enough to live in deep solutions. Its high fat composition also makes it more buoyant. Right: Dinoflagellates swim by means of a pair of flagella, which in this photograph can be faintly seen on the left. Members of the genus *Ceratium* also have long spines, probably to help thwart predators.

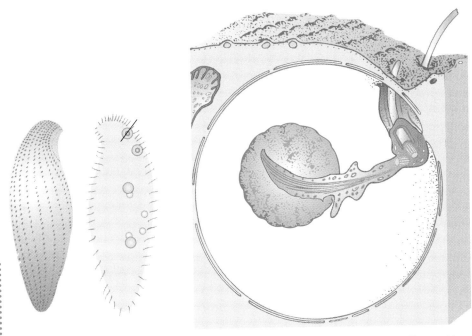

The structure of the presumed gravity sensor in the ciliate *Luxodes striatus*. The position of two of these Müller bodies in an *L. striatus* cell is shown at the left; the line through the upper Müller body indicates the plane through which the close-up of a Müller body on the right is viewed. These Müller vesicles, found only in this genus and *Romanella*, are composed of a ball of mineral granules suspended on a stalk in a water-containing sphere, 7 μm in diameter. Gravity is presumed to pull on the dense ball and bend the stalk, which is connected to cilia and may alter their pattern of beating or change membrane electrical potential. Depending on species, 1 to 30 Müller vesicles may be present in each individual.

The first convincing evidence that a microbe could actually sense gravity was reported only about ten years ago by Tom Fenchel of Denmark and Bland Finalay of the United Kingdom. The unusual ciliates of the genera *Remanella* and *Loxodes* are found primarily in sediments. *Remanella* exist in great numbers in the sandy sediments of marine environments but are restricted to a layer beneath the aerobic surface layer and above the anaerobic, sulfide-containing layer. *Loxodes* usually live in the freshwater sediment of lakes and ponds. When water near the bottom becomes anaerobic, these ciliates may be found several meters up the water column at the level where oxygen becomes available.

From detailed experiments carried out on *Loxodes striatus,* we know that under very low oxygen conditions, the cells swim mostly upward, while in water saturated with air, they swim mostly downward. In the key experiment, researchers placed cells in glass capillary tubes completely filled with water, some with oxygen and some without, and observed the behavior of individual cells after the tube (and cell) was inverted.

Whether cells were initially swimming upward in the absence of oxygen or downward in its presence, upon inversion the cells initiated a maneuver that turned them so they would continue swimming in the original direction.

If the cells were oriented passively by the force of gravity, one would predict that they would gradually turn to the preferred direction. Instead, the cells initiated a series of reversals in the beating direction of the cilia, causing the cells to rotate until they arrived at the appropriate orientation. This observation is consistent with the hypothesis that the cell senses its orientation to gravity. The question then becomes: How is it able to do so?

In all cases where we understand how gravity is sensed, and these cases all involve large animals, the organisms rely on special gravity-sensing organs. In these organs, a dense object such as a sand grain or a mineralized concretion sinks in the direction of gravitational pull and deflects the cilia of hair cells, which send signals to the brain. The gravitational sense organs of plants are also assumed to be based on dense particles, although their identity has not been established.

Romanella and *Loxodes* have a peculiar structure called the Müller body that might function as a gravity sensor. It is a spherical structure about 3 μm in diameter consisting of about 100 small mineral particles aggregated in an organic matrix. These bodies have high concentrations of unusual metallic elements: barium in *Loxodes* and strontium in *Remanella*. The metals are probably present in the mineral particles as the insoluble sulfate salts, which have a density four times that of water. The Müller body is suspended on a stalk in a vesicle about 7 μm in diameter, and the stalk is connected to a cilium. Significantly, a cilium is found to be a component of nearly all sensory cells in multicellular animals and is thought to initiate an electrical signal that spreads along the surface of the cell just like the signals in neuron cells. Thus, the structure of the Müller bodies is suggestive of an organelle that functions to detect the direction of gravity, and there is support for the hypothesis that these ciliates have a true sense of gravity.

Magnetic Bacteria

In the early 1970s, Richard Blakemore of the Woods Hole Oceanographic Institution discovered that certain bacteria swim in the direction of the earth's magnetic field. Analysis has revealed that these bacteria, termed "magnetotactic" by Blakemore, contain iron deposits that act as magnets. Because bacteria are very small, it does not take much force to rotate them, and it seems the earth's magnetic force acts directly to turn the bacteria into alignment with its magnetic field, just as it turns a much larger compass needle. Indeed, dead bacteria orient just as well as live ones. Therefore, this behavior is not a stimulus-response reflex but the action of a physical force; it is more analogous to falling under the force of gravity than to responding to its stimulation.

Magnetic bacteria are found in the sediments of many aquatic habitats and live in the absence of oxygen. They owe their magnetic properties to single-crystal particles of magnetite (Fe_2O_3) surrounded by a membrane. The magnetite particles in bacteria are just the right size to spontaneously form permanent magnets. In most bacteria, about 20 particles line up head to tail to form a straight chain, extending along the axis of the cell, that acts as a single magnet. In the earth's magnetic field, a chain of 20 particles generates enough force to keep the chain oriented, against the jostling of Brownian motion, within about 25° of the magnetic field direction most of the time.

One advantage of the magnets is that they reduce Brownian motion and keep the bacteria moving in a straight line, although the line of travel must of course

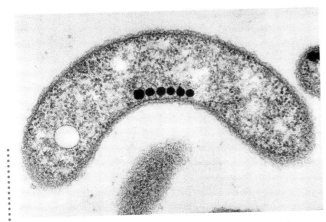

This electron micrograph of a thin section through a cell of *Aquaspirillum magnetotacticum* shows a row of six magnetic particles that orient the cell to the earth's magnetic field.

be parallel to the earth's magnetic field. At the predicted degree of orientation, swimming bacteria progress along the direction of the earth's magnetic field at an average speed that is at least 90 percent of their forward speed of locomotion. Without such alignment, their path would be a random walk, and their net progress through the environment would be much slower. Their adherence to the magnetic field lines works in their favor as long as the important environmental gradients are more or less parallel to the direction of the magnetic field, which is vertical at the earth's magnetic poles and horizontal near the equator.

In sediments, where oxygen diffuses only slowly from the overlaying water, the environment near the surface often changes dramatically over depths of a few millimeters or centimeters. Thus, it is particularly easy to see that the magnets would be useful to bacteria living at high latitudes but not near the equator. However, magnetic bacteria have been found at all latitudes. Under laboratory conditions, bacteria from northern latitudes swim predominately toward the north pole, which carries them to greater depths in northern latitudes, and bacteria from southern latitudes swim predominately toward the south pole, which in southern latitudes also carries them downward. Magnetic bacteria from the equator are equally likely to swim in either direction, north or south. The existence of magnetic bacteria at the equator can be explained by supposing that there are horizontal gradients as well as vertical ones. For instance, sediment particles of one chemical composition may be surrounded by concentric layers of differing composition, so that there are chemical gradients oriented in all directions. In this case, the advantage of moving in straight lines may outweigh the fact that the important gradients are parallel to the magnetic field direction only some of the time.

Navigating in the Darkness of Soil

The root-knot nematodes are troublesome pests from a farmer's point of view, because they damage many kinds of plants and cause billions of dollars worth of crop losses each year. Adult females of these nematodes live immobile in the root of a host plant, where they induce the plant to form giant cells, from which the nematode feeds. This largess is denied their own offspring, however, since the eggs they produce are extruded to the outside of the root. When the juvenile nematodes hatch, they do not feed but rely on stored energy supplies to support them while they search for an appropriate host plant. If they do not find a suitable host root before their energy stores run out, they simply die. Thus, we can expect that the juveniles are highly specialized for locating host roots and that they probably have interesting sensory capabilities to aid them in this search.

My lab conducted an extensive search for chemical stimuli released by roots that could guide juveniles of the southern root-knot nematode *Meloidogyne incog-*

Root knots formed by a root-knot nematode.

nita toward a host. We concluded that the only attractant chemical released by plants was carbon dioxide, and it is the main stimulus the nematodes use to move in on a particular plant. During this search, we stumbled on an interesting finding. For one of our assays, we had put a sample of root exudate at one end of a slab of agar resting in a small rectangular plastic box and an equal volume of plain water at the other end. After allowing several hours for the chemicals to diffuse through the agar, we placed a few hundred nematodes on the center of the agar slab. The boxes rested for two hours in a dark incubator that had been advertised as maintaining a constant temperature; we then examined the boxes under a microscope to see if the nematodes had moved toward or away from the end containing the root sample. We were careful to include, as controls, some boxes in which we had placed plain water at both ends.

Surprisingly, we often found that even in these control experiments, significantly more nematodes moved toward one end than the other. After exploring a number of possibilities, we determined that the nematodes were responding to temperature gradients. Our "constant temperature" incubator had about a

one-degree difference between the front and the back, and the nematodes seemed to respond to this small difference.

Now this seemed to be a very small temperature difference—too small for us to feel—and suggested that the nematodes might be specialized to respond to small temperature gradients for some reason. Indeed, some investigators had already reported finding a sensitive response to subtle temperature gradients in other plant-parasitic nematodes, and they had suggested that nematodes might use the heat generated by the metabolism of plant roots to find the roots. Calculation suggests that the temperature gradients from a root would be large enough to be theoretically detectable, but they are much smaller than the gradients in soil produced by other sources. The signal from the roots would be completely submerged in noise from other sources of heat.

In my lab we decided to characterize the behavior of the nematodes to temperature gradients—in part out of curiosity and in part so that we could prevent this response from interfering with our chemical assays. The method we chose was similar to the method we were already using in our search for chemical stimuli. We placed the nematodes at the center of a slab of agar contained in a small rectangular plastic box. Then we placed the box in an aluminum channel, the ends of which were in contact with water baths at different temperatures. The temperatures had been adjusted to provide a gradient with a steepness of about 0.1 °C/cm, within a range that we could vary when we repeated the experiment. The nematodes were allowed to creep around for two hours, and afterward we recorded their distribution along the agar. We found that the direction of migration depended on the average temperature. At extremely low or high temperatures, the nematodes did not move at all, immobilized by heat or cold stress. At temperatures just above the lower limit of migration, the nematodes migrated toward warmer temperatures. Conversely, at temperatures just below the upper limit, the nematodes migrated toward cooler temperatures. Overall they acted as if they had a preferred temperature and moved toward it from both lower and higher temperatures.

This kind of behavior has been observed in many organisms, and it is usually assumed that the organism prefers the temperature to which it is physiologically best adapted. In some cases, the preferred temperature and the "best" temperature can be shifted by holding the organism at a new temperature (the "acclimation" temperature) for several hours or days during which the organism acclimates to the new temperature. When we held nematodes at altered temperatures for several days, we found that the preferred temperature was indeed shifted.

However, we also observed something very surprising. The preferred temperature did not coincide

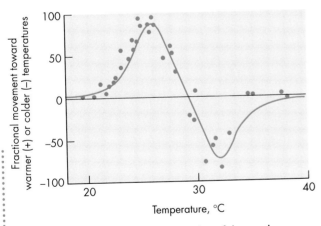

The migration of infective juveniles of the southern root-knot nematode *Meloidogyne incognita* along a temperature gradient. Groups of nematodes placed at temperatures below their "preferred" temperature of 29 °C migrate toward higher temperatures, whereas groups placed at temperatures above the optimum migrate toward lower temperatures.

with the acclimation temperature but was, in all cases, several degrees higher. This observation was unprecedented, and no explanation was immediately obvious.

Our next basic question was: How sensitive are these animals? We studied this question in two ways. The straightforward approach was to make temperature gradients more and more shallow until the nematodes no longer responded. The results showed that the nematodes could respond to gradients as shallow as 0.001 °C/cm. This is the most sensitive response to temperature gradients that has been determined for any organism. Such small gradients are difficult to eliminate even in the laboratory. In nature, the nematodes would almost always be exposed to a gradient that they could follow.

We also looked at the nematodes' response to temperature over time using a computerized tracking technique that I had developed for studying how nematodes respond to chemicals. In this technique, the nematodes were observed by a video camera connected to a computer. We placed the nematodes on a thin slab of agar that was illuminated by oblique rays of light that were too oblique to enter the camera, so the instrument saw only light that was scattered. Since the nematodes scatter much more light than the agar, they showed up as bright images against a dark background. The computer was programmed to look for bright objects and keep track of their location. Every three seconds it checked every recorded location to see if the location was still bright. If not, it initiated a search for the nearest bright position. This procedure caused the computer to follow the trailing edge of each nematode. A modest computer can follow several hundred nematodes simultaneously and record their average rate of locomotion in real time.

This average rate is influenced by their responses to stimuli of many kinds. Under constant conditions, the nematodes sometimes pause or back up, behaviors

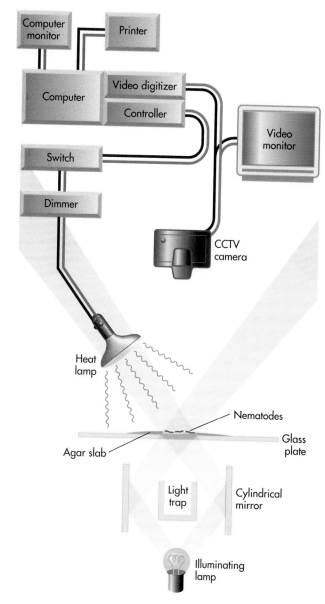

Computer monitor

Printer

Computer

Video digitizer

Controller

Video monitor

Switch

Dimmer

CCTV camera

Heat lamp

Nematodes

Agar slab

Glass plate

Light trap

Cylindrical mirror

Illuminating lamp

The apparatus depicted here was used to record nematode responses to very small changes in temperature. The video camera viewed the nematodes as bright sources of scattered light. The computer converted the video signal to digital form; a program followed the changing positions of about 200 nematodes as they crawled around on the agar, and saved the average rate of locomotion. The computer also controlled a dim heat lamp that heated the agar at a rate of about 10^{-4} °C/s; the agar cooled at about the same rate when the lamp was turned off.

and the nematodes travel at a faster average rate. Conversely, if stimuli become less favorable, the nematodes pause and back up more frequently, and their average rate of locomotion decreases.

In order to change temperatures, the computer controlled a dim incandescent lamp that illuminated the agar and the nematodes. After the lamp was turned on or off, the temperature of the agar initially changed at a rate on the order of 10^{-4} °C/s. Like other microbes, the nematodes are so small that the rapid diffusion of heat keeps them at essentially the same temperature as the agar.

The nematodes responded within a few seconds. To react this quickly, they must be able to detect a total change of less than 0.001 °C. We can be sure that the stimulus they responded to was temperature rather than light or some other factor, because their rate of locomotion increased or decreased according to whether the ambient temperature was above or below their preferred temperature. It is very unlikely that some other stimulus (such as light) would have such a peculiar temperature dependence.

When the temperature moved away from the preferred temperature of 26 °C, the average rate of locomotion decreased, as expected, though only until the nematodes adapted, within half a minute. When the

that reduce the average rate of locomotion that is recorded. If stimuli become more favorable (perhaps the temperature moves closer to the preferred temperature), the pauses and reversals become less frequent,

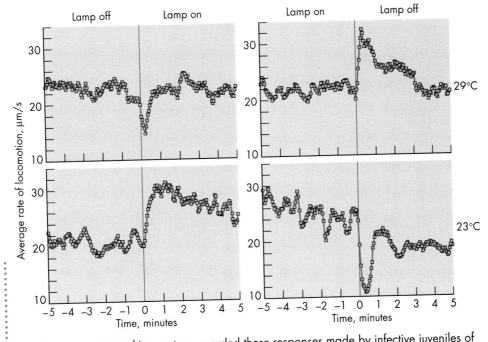

Our computer tracking system recorded these responses made by infective juveniles of the nematode *M. incognita* to small changes in temperature. The nematodes had been kept at 23 °C and had a preferred temperature near 26 °C; they were tested at ambient temperatures above (29 °C) and below (23 °C) the preferred temperature. When the heat source was turned on or off at time zero, the average rate of locomotion changed within a few seconds, speeding up if the temperature changed toward the preferred temperature and slowing down when the temperature changed away from the preferred temperature.

temperature moved toward the preferred temperature, the average rate of locomotion increased, as expected, but in this case the nematodes took several minutes to adapt. The differences in the rates of adaptation are another example of the behavior, earlier described on page 71, that leads nematodes to move more efficiently along a stimulus gradient.

These interesting observations led us to ask why the nematodes respond in this way. Is their behavior an accidental consequence of their physiological mechanisms, or does it serve a function for the nematodes?

We needed to understand how these responses cause the nematodes to move in their natural environment.

There are few sensory stimuli that vary with depth in a reliable fashion, but one that we know about is the amplitude of daily temperature variation. Sunlight falling on land surfaces causes relatively large changes in temperature. Daily changes of tens of degrees in the air and soil close to the surface are common as the soil is heated during the day and radiates heat to the sky at night. Heat travels through soil primarily by conduction, and in most cases the flow of heat and resulting temperature can be accurately estimated by calculating the conduction of heat through a uniform solid. Such calculations suggest that within 30 cm of the surface, where root-knot nematodes are to be found, the temperature gradients are determined primarily by daily variations. The largest gradients are usually present near the surface, with 1 °C/cm being common in the top few centimeters. The overall pattern can be described as a temperature wave, varying in time with a period of 24 hours, that propagates down into the soil at a rate of about 30 mm/h. This wave has a maximum amplitude of temperature variation at the surface, and the amplitude declines by half with every 70 mm of additional depth.

The very complex environment of the soil, which contains many chemical stimuli that might interfere with the response to temperature, makes studies in soil difficult, so I turned to computer modeling. Since a fairly accurate representation of the temperature gradients in soil can be calculated, it only remains to describe the behavior of the nematodes in computer language, place them in the temperature pattern of soil, and ask how they move.

In working this out, we realized that we needed two additional pieces of information: (1) a more accurate estimate of the rate of acclimation and (2) the temperature range within which the nematodes would remain active. For technical reasons the rate of acclimation was difficult to determine for the root-knot nematodes, but fortunately I had data on both these points for the free-living nematode *Caenorhabditis elegans,* which also lives in soil but feeds on bacteria. The thermal acclimation of this nematode is half complete in two hours. As for the temperature range of locomotion, the nematodes become inactive when the ambient temperature is more than 6 °C from the optimum temperature.

The computer model predicts that a nematode in

The pattern of temperature changes in soil, covering a period of 96 hours moving from left to right. Daily temperature changes are large at the soil surface (top), cycling from warm (red), through average (yellow), to cool (blue) and back. Moving downward into the soil toward a depth of 20 cm (bottom), the temperature range decreases (more yellow), and the pattern of change shifts to later times (rightward).

a temperature pattern typical of soil will move upward during part of the day and downward during other times. However, the total distance moved upward will generally differ from the distance moved downward, and there will be a net drift of the nematodes upward or downward. Thus, following a temperature gradient can lead to a net change in depth even though the average temperature is the same everywhere. Specifically, nematodes starting within the top 15 cm of the soil drift toward a depth of about 5 cm, while nematodes starting deeper than 15 cm drift deeper.

Forrest Robinson of the U.S. Department of Agriculture has come closest to recording how nematodes actually move in their natural environment. Robinson filled plastic tubes with damp sand and buried them in a sand-filled box containing two automobile transmission coolers. The temperature of the water pumped through the two transmission coolers was altered on a set schedule to generate temperature gradients similar to those measured in late summer in a cotton field in eastern Texas. Robinson then injected nematodes into the center of the tubes. After either 24 or 48 hours, he extruded the sand and cut it into sections, then determined the number of nematodes in each section. He found that *M. incognita* moved in the direction of the temperature found at 5 cm depth and away from the temperature found at 25 cm depth, and did so independent of the tube's actual orientation with respect to gravity.

What function could this behavior serve? Most nematodes probably hatch within the top 15 cm of soil,

and the 5 cm depth might well be the optimal depth at which to search for the roots of an appropriate host. At shallower depths, the nematode runs more risk of heat damage or dehydration; at greater depths fewer roots may be available for invasion. Thus, the nematodes would be guided by thermal gradients until they get sufficiently close to a root to detect chemical gradients, probably of carbon dioxide. Although the pattern of movement followed by nematodes is not very efficient, this behavior allows a simple animal to make use of a complex pattern of stimuli to find a particular depth. No other stimulus is known that could be used by the nematodes to solve this problem.

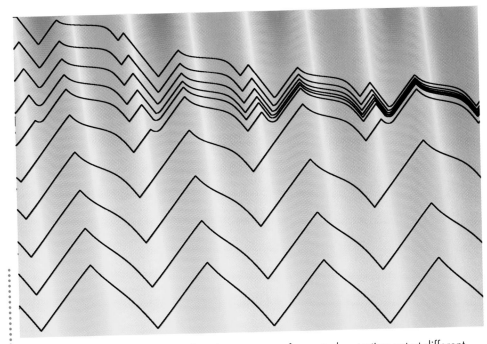

The black trails show the predicted movements of nematodes starting out at different depths, superimposed on the temperature pattern from the facing page. Those starting out within the top 15 cm drift toward a depth of about 5 cm, which may be the optimal depth at which to search for the roots of a suitable host plant. Those that happen to start out deeper than 15 cm drift deeper, but probably relatively few nematodes hatch at that depth.

Slime Molds Navigating in Soil

The cellular slime mold *Dictyostelium discoideum,* which like root-knot nematodes lives in soil, presents a different puzzle concerning a response to temperature gradients. Amoeboid cells of this slime mold grow by feeding on bacteria in soil. When a dense population experiences a decline in food availability, the cells aggregate into masses of cells called "slugs." For a few hours, the slugs migrate over distances of a few centimeters, guided by light and temperature. They then form a stalk that lifts spores away from the surface for more efficient dispersal.

These slugs are the most sensitive to temperatures of any reported organism aside from nematodes and orient to gradients as small as 0.05 °C/cm. In this case, the direction of movement is *away* from a temperature that is about 2 °C *below* the acclimation temperature. To avoid a temperature so near the acclimation temperature seems peculiar. Organisms would of course avoid temperature extremes, which are invariably detrimental, but why should they avoid a temperature near the middle of their temperature range? I investigated this question by using a computer model similar to the one I had used for nematodes.

An organism that avoids the mean temperature will move toward higher temperatures when the temperature is above average and toward lower temperatures when the temperature is below average. At a given location, the temperature is above average for half of each period and below average for the other half. Similarly, the spatial gradient of temperature is directed upward during half the cycle and downward during the other half. However, these two cycles are out of phase by $\frac{3}{8}$ cycle. An organism avoiding the mean temperature will move upward when the temperature is above average and the gradient is directed upward, but it will also move upward when the temperature is below average and the gradient is directed downward. Upward movement will take place during two intervals in the cycle—during the first $\frac{3}{4}$ of the warm half and during the first $\frac{3}{4}$ of the cool half. Thus, over the long term, movement will be upward three-quarters of the time and downward one-quarter. If the rate of locomotion is equal in both directions,

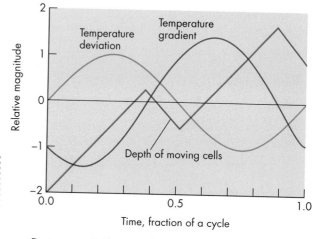

During one 24-hour cycle, an organism (green line) avoiding the mean temperature, like *Dictyostelium* slugs, moves upward during two parts of the cycle (when the temperature is above average and the gradient is negative or when the temperature is below average and the gradient is positive) and downward during intervening parts of the cycle (when the temperature is above average and the gradient is positive or when the temperature is below average and the gradient is negative). Note that the two curves indicating temperature deviation from the average (red) and the direction and strength of the temperature gradient (blue) are displaced from one another. (Positive values represent warmer temperatures toward the surface.) Because of the size of the shift in the two curves, upward movements occur for three-quarters of the cycle and downward movements for only one-quarter. Consequently, the organism drifts toward the surface.

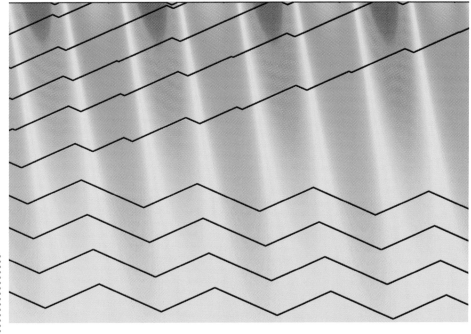

The black trails, superimposed on the temperature pattern (colors) in the top 30 cm of soil, show the paths of organisms that, like slugs of the slime mold *Dictyostelium*, avoid a temperature offset below their acclimation temperature. Individuals starting out within the top 20 cm of the surface move toward the surface, while those that start out at a greater depth move deeper.

there will be a net movement toward the surface at $\frac{1}{2}$ ($= \frac{3}{4} - \frac{1}{4}$) the rate of locomotion.

The balance of upward and downward movements is modified if the organism moves at rates comparable to the rate of penetration to the thermal wave (2 to 3 cm/hr) or the organism acclimates to new temperatures. If the rate of locomotion is as large as the speed at which temperature penetrates the soil, the organism will simply follow the preferred temperature down into the soil. At the slime mold's observed rates of locomotion and acclimation, the computer model predicts that the slime mold would move to the surface from a depth of 10 cm in 101 hours. In reality, the slug stage lasts only a short time since the slug does not feed. It is reported to persist as long as 24 hours, and during this time the slug moves a distance of about 5 cm. Consequently, it could reach the surface from a depth of 2 to 3 cm. It seems obvious that the slug benefits by moving to the surface in order to form a fruiting body at a location from which spores could be released into the open air for dispersal.

It is striking that even an organism as simple as the slime mold can make use of the information in the complex pattern of temperature changes in soil to guide its locomotion to the surface. Many other microbes will probably be found to make use of subtle stimuli in much more sophisticated ways than have previously been appreciated.

Some fungi obtain nutrients by trapping animals that happen by. The fungus *Arthrobotrys anchonia,* whose slender strands are seen in this scanning electron micrograph, forms rings that suddenly constrict after being touched on the inside, an efficient mechanism for trapping a nematode.

CHAPTER **6**

Stationary Feeders

When nutrients are scarce, *Euglena* and *Paramecium* can swim to a place where they are more plentiful. Amoebae can creep along until they find bacteria or other edible cells. Many other organisms are stationary, however—the fungus growing on a tree trunk, the bacteria in a root nodule. These organisms need strategies for making nutrients more available at the location where they happen to be. Some release digestive enzymes that degrade nearby organic matter, freeing nutritive molecules that can diffuse back to the organisms. Others release toxic chemicals that cause the cells of other organisms to become leaky or even break open, liberating their nutrients. In some cases, two types of organisms form a special association and exchange nutrients with each other—each organism contributing a different kind of nutrient.

All these strategies rely on the fact that diffusion dominates molecular transport on the micro scale. On the macro scale of large organisms, currents in the environment would carry released enzymes, toxins, and nutrients downstream, where they would act on other organisms or material; then the liberated nutrients would be carried further downstream—away from the organisms that had invested in producing and releasing the enzymes, toxins, or nutrients in the first place.

On the micro scale, in contrast, molecules diffuse equally in all directions after their release and do not move very far away. Thus, any liberated nutrients have a high probability of being captured by the organism that originally invested in the molecules that achieved the liberation. The organism may even be equipped with cellular uptake systems that can deplete the returning nutrient from the surrounding area and create a concentration gradient that enhances diffusion to the organism.

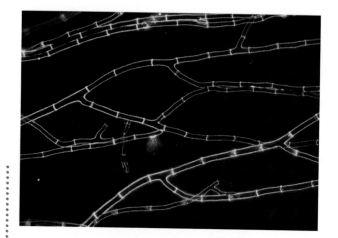

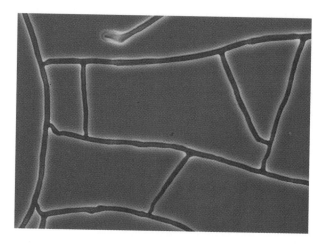

Left: This darkfield light micrograph clearly shows the cell walls of a fungus, *Sordaria fimicola*. The long branching strands, or hyphae, compose its mycelium. This fungus, like some, forms crosswalls, called septa, that divide hyphal strands into separate compartments. Right: The hyphae of some fungi fuse to one another to form an interconnected network. This interconnectedness can provide a number of advantages, including more flexible movement of nutrients through the mycelium, added strength for structures such as mushroom stalks, new mechanisms for genetic combination, and new mechanisms for trapping animals.

External Digestion

With their tough cell walls of chitin, fungi cannot engulf food particles and digest them internally as do many protozoa. They can only absorb small molecules dissolved in water or small pieces of protein degraded to a length of less than six amino acids. Although unable to travel to new locations in search of food, fungi do not wait passively for small molecules to appear nearby; rather, they excrete enzymes into their environment that break down large and insoluble molecules into small parts that can be absorbed.

One nutritious molecule that is small enough for fungi to absorb is glucose, the "blood sugar" of mammals. But glucose is the favorite carbon and energy source of most aerobic organisms and doesn't last long in the natural environment. Fungi often obtain their own supply of glucose by secreting enzymes that degrade the most abundant of all organic molecules, cellulose, which is formed from chains of glucose molecules. Fungi are one of the few kinds of organisms able to liberate glucose from cellulose, since they are one of the few equipped with the variety of powerful and uncommon enzymes that must act in concert to break apart this firmly linked molecule. This method of obtaining food has allowed fungi to prosper in virtually all habitats and has established the great ecological importance of fungi in the recycling of nutrients. In addition, their ability to secrete highly stable enzymes has been exploited commercially. Many soft drinks and canned fruits, for example, are sweetened by high-fructose corn syrup that has been formed by the breakdown of plant starch and the conversion of its glucose to the sweeter fructose, all accomplished by enzymes from fungi.

Fungi do not secrete their valuable and costly enzymes willy-nilly, but only when appropriate. Usually a fungus holds back its production of cellulose-degrading enzymes unless a source of cellulose is actually nearby and the organism is in need of its component glucose. Where glucose is plentiful, a fungus stops synthesizing enzymes for digesting cellulose and other glucose-containing substrates, such as starch, and even stops synthesizing uptake systems for other small sugars. Similarly, fungi secrete protein-degrading enzymes only when protein is present *and* the organism is starved for carbon, nitrogen, or sulfur—all of which a protein can provide.

The second most abundant organic material on earth after cellulose is lignin, a tough substance that impregnates wood and helps protect cellulose from attack. Lignin is a complex, ill-defined material, and fungi are the only organisms we know of that degrade it completely. In contrast to the way they attack most substrates, fungi do not break lignin into small molecules that can be absorbed for food. Rather, they use hydrogen peroxide to oxidize lignin in place. The broken down and solubilized lignin diffuses away, exposing the cellulose to enzymatic attack. Although the fungi do not gain carbon or energy by destroying lignin, they gain access to the cellulose and other carbohydrates of the wood.

Fungi grow in the form of a finely branched network, or mycelium, that can release digestive enzymes and take up nutrients at many sites distributed over a large volume. Nutrients that happen to diffuse away from the site of enzyme release, and might otherwise be picked up by competing organisms, are likely to be captured at other uptake sites belonging to the same organism. The hyphae of fungi seem to grow and branch in patterns that are optimal for exploring the environment and obtaining nutrient resources, although the mechanisms used for doing this are poorly understood. The fungus *Neurospora crassa* alters its branching angle as the mycelium grows—from 90° at the start of growth down to 60°. The smaller angle is thought to direct new growth away from the already exploited center of a well-developed mycelium. In many cases, there is also evidence that a growing hypha tends to turn away when it approaches another to within 10 to 20 μm. This behavior helps keep hyphae spread out over the environment and not competing for the same nutrients.

On the other hand, the hyphae of certain fungi sometimes fuse together, after guided growth toward one another from a distance of 10 to 15 μm. The fusion of hyphae converts the mycelium from a branched structure to an interconnected network offering alternative pathways through which the fungus can transport nutrients. If the hyphae were not fused, the destruction of a single segment would divide the mycelium into two separate organisms that could no longer exchange resources.

Whereas some fungi may send their hyphae out blindly, albeit in an optimal pattern, many seem to know where they are going. The fungus *Phanerochaete velutina,* which causes the decay of wood on a forest

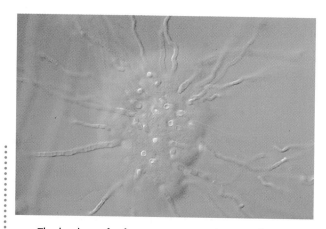

The hyphae of a fungus grow straight toward a colony of bacteria, the fuzzy blue patch in the center.

floor, is able to grow toward "fresh" wood detected from a distance. After thoroughly exploiting a piece of wood, it sends out a few exploratory hyphae in all directions through the surrounding soil. If one hypha approaches uncolonized wood, it grows *directly* toward it, probably guided by volatile chemicals that emanate from the wood, and eventually produces a dense mycelium over the new wood mass.

In an extensive survey of 100 fungal species, George Barron, of the University of Guelph in Ontario, found four species able to initiate hyphal branches that grow straight toward colonies of the soil bacteria *Agrobacterium tumefaciens* or *Pseudomonas putida* living within a few hundred micrometers. Once within the colony, the hyphae branch extensively, and the bacteria disappear, presumably having been digested and their nutrients absorbed by the hyphae. After no bacteria are left, the induced hyphae are emptied, and their nutrients probably transported back to the main hyphal mass. The four species that exhibit this behavior are known for growing on the difficult-to-digest lignin and cellulose components of plants. As food for growth, these materials are deficient in nitrogen-containing molecules necessary for synthesizing proteins and nucleic acids. Consequently, nitrogen-rich bacteria represent a particularly valuable resource, and it is sensible that these fungi have developed the means to capture nutrients from any bacteria growing nearby.

The fungus's strategy of sending out filaments in search of nutrients is shared by other stationary organisms, especially the roots of plants. Some nutrients that plants require, including phosphate and ammonia, bind to soil particles and are rendered relatively immobile. Consequently, roots create a zone of depletion around themselves as they take up nutrients from the soil. To obtain more nutrients, the root must grow

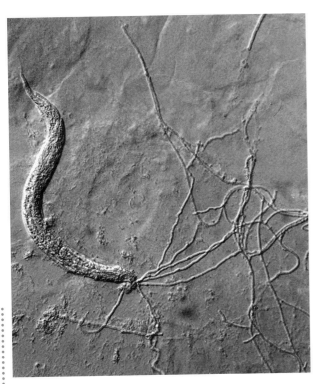

The hyphae of a fungus grow straight toward the mouth of a nematode that the fungus has paralyzed by secreting toxins. Nematodes have a very tough cuticle, and the mouth provides an easy point of entry.

Setting Traps

To fungi that grow on rotting wood and other substrates low in available nitrogen, passing animals also represent a rich source of nutrients—probably even more valuable than the colonies of bacteria just mentioned. Many of these species have developed adaptations for snaring nematodes, bdelloid rotifers, and even small crustaceans such as springtails and copepods. When the fungus senses a suitable animal in the vicinity, perhaps by recognizing specific waste products or carbon dioxide, it forms specialized adhesive knobs, branches, or nets and holds these erect from the substrate to increase the chance of an animal contacting one of these protruding structures and becoming stuck.

Primitive fungi, which are missing the crosswalls called septa, form extensive hyphae that lack the specialized knobs and other structures formed by many more advanced fungi; to capture nematodes they rely solely on an adhesive that covers their surface. This adhesive seems to be quite strong: even though large nematodes that are able to push against surface tension can usually break free of a single point of attachment, the thrashing of a nematode caught at one point usually causes it to be caught at other points. For some reason, the adhesives appear to capture only certain types of organisms. Protozoa escape untouched, and even among the nematodes only certain species are vulnerable to the fungus' trap.

Certain more complex fungi produce three-cell rings, and nematodes of just the right thickness become caught when they chance to crawl through a ring. The most sophisticated trapping devices are constricting rings produced by species such as *Arthrobotrys anchonia* and *Dactylaria brochopaga*. When a nematode heads through a ring, it makes contact with

into areas of soil not yet depleted. A plant maximizes its uptake by having many rapidly growing roots of small diameter, and plants often form root hairs less than 10 μm in diameter.

Another solution is for the plant to form an association with a fungus whose hyphae provide an even more efficient absorptive structure. Most vascular plants can form such associations, called mycorrhizae. The fungus provides scarce nutrients from the soil to the plant, and the plant provides carbohydrate food to the fungus, which depends on the plant for energy.

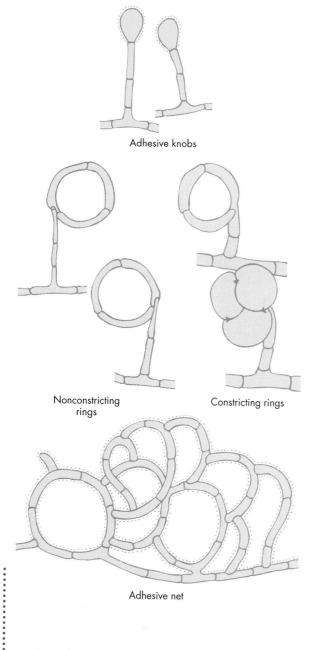

Adhesive knobs

Nonconstricting
rings

Constricting rings

Adhesive net

Fungi have invented a variety of knobs and rings for trapping nematodes, some of them illustrated here. The dotted surfaces indicate the presence of a substance that sticks to the cuticle of nematode prey.

the interior surface and triggers the cells to swell inward. An empty ring swells completely shut, and any nematode within the ring is held tightly in the noose. There is a delay of a few seconds between the moment of contact and the initiation of swelling, perhaps to let the nematode get farther into the ring. But once started, the ring completes swelling in about a tenth of a second, giving the nematode little time to react. In many cases, the struggles of a nematode caught in one ring only cause it to become caught in another ring, and it is then firmly trapped. (See page 110.)

After catching a nematode, the fungus sends out a specialized "infection" hypha that penetrates the cuticle and sends branches throughout the nematode's body. Once the fungus has absorbed all the nutrients that the nematode has to offer, the cytoplasm of the hyphae flows out of the empty carcass and back to the main hyphal mass.

The trapping structures of many fungi are only weakly attached to the main hyphae, and the nematode may break one off and carry it away, attached to its body. But the nematode has not escaped. The broken-off piece is usually a viable fragment of hypha that eventually infects the animal. Thus the trapping devices are also an effective means of dispersing the fungus, with the host nematode providing the means of locomotion.

George Barron has discovered a species of fungus that captures copepods, the first fungus ever observed to trap these tiny aquatic crustaceans. (Copepods have the same sort of articulated carapace and appendages as shrimp and other crustaceans, but may measure only a millimeter or so in length.) In 1990 Barron isolated a new fungal species, *Dactyella copepodii*, from rotted manure and leafy debris that had been gathered in Ontario. When exposed to air in the laboratory, the hyphae produce small fruiting stalks capped at the tip by a single propagule, the structure from which new

The fungus *Arthrobotrys oligospora* is depicted growing on an agar plate with three nematodes that it has trapped. The fungus forms the sticky rings only after it has detected the presence of nematodes, presumably by sensing some chemical the nematode has excreted. The three fruiting bodies that are also present carry spores away from the surface for better dispersal.

hyphae are produced. After being exposed to certain copepods, the propagules germinate and produce a few short hyphae with adhesive swellings at their tips. If a copepod happens by, the hyphae stick to it and the copepod carries the fungus away. In about a day the copepod will be dead, killed by the hyphae growing through the copepod's cuticle and feeding on its insides. Hyphae then grow out of the copepod's body

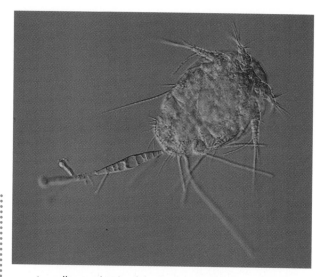

An adhesive knob of the fungus *Dactyella copepodii* has attached to the rear of a larval copepod (lower left). The fungus will soon send hyphae into the body of the copepod and digest it.

and produce more branches with the adhesive tips. If another copepod happens by, it may become stuck to the fungus, which is now too big for the copepod to carry away. The fungus may then grow even larger and capture more copepods, if they keep coming. Sometimes an adhesive branch may break off at a weak area near the base when the copepod tries to pull away. The escaping copepod carries a piece of hypha with it, and again its efforts only help disperse the fungus, as the hypha will soon infect the animal.

The unusual bacteria *Streptomyces,* which grow mostly in soil, resemble fungi in structure as well as in many of their habits. The bacterial cells form an extensively branched mycelium, usually divided into two types. The primary mycelium grows on or through dead vegetation or other organic matter on which it feeds, while a secondary mycelium grows into the air and eventually forms chains of spores. These bacteria are not fastidious in their nutrient requirements, and can flourish on inorganic nitrogen and without vitamins. They also grow in conditions that are too dry for other types of bacteria, although they do not like the additional carbon dioxide common in waterlogged soils.

Since mycelial growth is well suited to the strategy of obtaining nutrients by external digestion, it is not surprising that, like fungi, *Streptomyces* secrete enzymes into the environment. These enzymes are capable of digesting complex, recalcitrant plant and animal residues such as chitin (from fungal cell walls and arthropod exoskeletons), keratin (from hair and feathers), and lignocellulose (from wood). Altogether scientists have identified 97 species of bacteria belonging to 35 genera as specialized for degrading cellulose, some forming mycelia, and others existing as individual cells that attach to dead leaves, animal carcasses, or pieces of wood.

Humans and most other species benefit from the cellulose-digesting abilities of certain microbes for, once the glucose is liberated from cellulose and absorbed by the microbes, it becomes available to the microbes' predators and subsequently to others further up the food chain. Some animals, however, take advantage of these abilities in a way that is beneficial to animal and microbe both.

Termites and their Guests

Cellulose is difficult to break down—which is why it is a good building material for plants and humans. Nevertheless, termites and a few other animals are specialized to digest materials such as wood that are high in cellulose. Termites and some plant-eating mammals obtain nutrients from cellulose (which makes up 50 percent of dry plant mass) only by maintaining an intestinal compartment that is congenial to cellulose-degrading protozoa and bacteria.

The rumen of a cow that is eating grass contains several hundred species of bacteria (at 1 to 10 billion cells per cubic centimeter), 40 to 50 ciliate species (at 100,000 cells per cubic centimeter), and many fungi. The three groups are present in roughly equal biomass. Some of the bacteria and some of the ciliates degrade cellulose, and all of the fungi do. The different groups work in concert, the fungi breaking down larger food particles, exposing new surfaces to the bacteria, and the bacteria finishing the job and probably doing most of the work. At least some of the ciliates are probably ingesting bacteria rather than cellulose.

The so-called lower termites are able to prosper eating sound wood that has not been attacked by fungi. They and certain wood-eating roaches harbor several species of large flagellate protozoa in their

A living *Trichonympha campanula*, one of the many flagellate protozoa that digest cellulose and live symbiotically in the guts of termites. Its amoebalike, wide posterior end can extend pseudopods to engulf bits of wood that it encounters in its host's hindgut.

hindguts. The protozoa apparently ingest wood particles, aiding in their digestion. Bacteria resembling spirochetes often cover the surface of the protozoa; they apparently provide a means of locomotion to their protozoan hosts, which use their flagella only for steering. Termite colonies that have been deprived of their flagellates can survive on partially decomposed wood but not on the sound wood that normally supports their growth. In some cases, the flagellates make up more than a quarter of the termite's total body weight.

The specialized flagellates that live in termites have not been found elsewhere. The termites and flagellates have apparently evolved together—that would explain why related flagellates grow only in related termites. The communal life of termites probably helps them continue their dependence on the flagellates. The termites habitually regurgitate food to other members of their colony, and by this behavior can inoculate new members of the community with the microbes.

You might suppose that the termite takes much of the liberated glucose for its own use, but the microbes use most of the glucose for themselves. How, then, does the host benefit from the digestion of cellulose? The hindguts of termites and the rumen of cows, where cellulose is digested, are anaerobic—there is little oxygen present. Microbes that metabolize organic molecules in the absence of oxygen must excrete some organic waste product—as yeast produce alcohol. Thus, although the microbes use most of the glucose liberated from cellulose, they must excrete some organic molecule. For symbiotic intestinal microbes this molecule is acetate or another short-chain fatty acid, and these are absorbed by the intestine. The host animal, having access to oxygen from the air, can metabolize the acetate or fatty acid to carbon dioxide, obtaining energy for itself.

One problem of living on a diet of wood is that the material is very low in nitrogen. Like most animals, termites are about 10 percent nitrogen by dry mass. Fresh, green plant tissues and the cambium of trees are 1 to 5 percent nitrogen, but sapwood is only about 0.1 percent nitrogen, and heartwood has even less of the element. Thus termites that lived exclusively on sound wood would have to efficiently extract nitrogen from large volumes of wood if this were their only source of the nutrient. This problem puzzled biologists for many years. However, in 1973, investigators discovered that termites had a rare ability: they could take nitrogen from the air and convert it to a usable form. You won't be surprised to learn that it is presumably not the termites themselves that are doing this, but some of the bacteria in their intestines. In fact, it is safe to say that all life depends on this ability of certain bacteria.

Nitrogen-Fixing Plants

All cells require the element nitrogen to build proteins and nucleic acids. Although nitrogen is abundant in the atmosphere, the inorganic form (N_2) in which it occurs there is unusable by the vast majority of organisms. However, certain bacteria and cyanobacteria are capable of breaking up N_2 and producing forms of nitrogen (NH_3, NO_2^-, and NO_3^-) usable by most organisms. Thus, the great majority of organisms are dependent on these few "nitrogen-fixing" prokaryotes.

All organisms capable of carrying out nitrogen fixation employ a similar mechanism, carried out by similar enzymes, and it probably evolved only once, very early in the history of life. The reaction consumes much energy—to fix one molecule of N_2 requires about 25 molecules of ATP, or most of the energy from oxidizing one glucose molecule. Consequently,

an organism will not spend the energy to fix N_2 if other forms of nitrogen are available. Aside from the energy cost, a nitrogen-fixing organism faces another difficulty: one of the enzymes necessary for nitrogen fixation is destroyed by oxygen, and yet oxygen is required for efficient formation of ATP. Because of the conflicting priorities—to have oxygen and to not have oxygen—nitrogen fixation is often carried out in a symbiotic relationship between a nitrogen-fixing prokaryote and another organism. The nitrogen-fixer trades nitrogen in the form of ammonia (NH_3) to its partner in exchange for energy-providing carbohydrates. This partnership allows many eukaryotes to grow in nitrogen-poor environments that they would otherwise find uninhabitable.

The eukaryotes that benefit from these symbioses include marine diatoms, the fungi of certain lichens, shipworms, termites, and certain plants. Because of its importance to agriculture, the association between rhizobium bacteria and leguminous plants has been studied extensively. The plant devotes 10 to 30 percent of all it produces by photosynthesis (in the form of the sugar sucrose) to the root nodules harboring the bacteria. The nodules convert the sucrose to the four-carbon compound malate, which is transported out of the plant cell and provided to the nitrogen-fixing bacteria. In exchange, the bacteria release a substantial proportion of the synthesized NH_3 to the plant.

Only certain kinds of plants develop the close association and specialized root nodules required for nitrogen-fixing bacteria. The proper match of plant and bacteria is orchestrated through the release and sensing of chemicals. The plant root synthesizes and releases a certain flavonoid compound. The flavonoid binds to a receptor protein in a compatible rhizobium, triggering the activation of genes in the bacterial cell. Some of these activated genes cause the formation of a specific carbohydrate. Compatible plants recognize this chemical and initiate changes in the root to form a nodule incorporating the bacteria. Thus, one can think of the formation of a nitrogen-fixing nodule as the result of a chemical conversation between the plant and an appropriate bacterium.

One type of microbe can often acquire resources that another cannot, so two types that form an association may compete for resources more successfully than either could alone. Many types of eukaryotic microbes that do not themselves carry out photosynthesis harbor certain kinds of algae that do. The algae

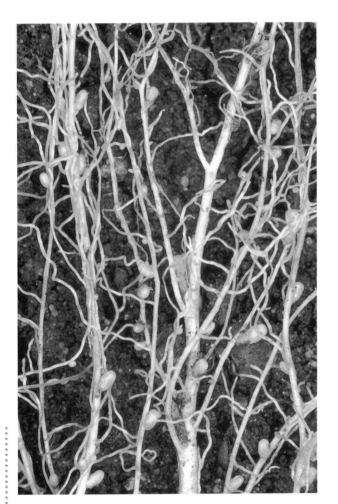
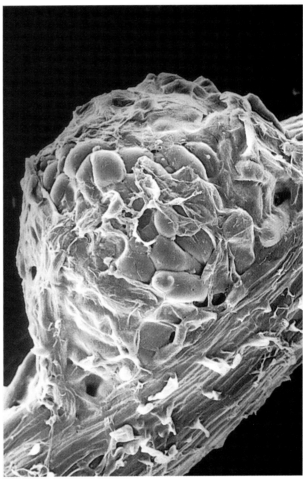

Left: Nitrogen-fixing nodules on the roots of white clover, formed by the nitrogen-fixing bacteria *Rhizobium trifolii*. Because of this nitrogen-fixing ability, farmers often grow a crop of clover to add nitrogen to the soil. Right: A scanning electron micrograph of a nodule on the root of a pea plant, caused by the nitrogen-fixing bacterium *Rhizobium leguminosarum*. This association allows pea plants and other legumes to live in soils that would not otherwise have enough nitrogen to support their growth.

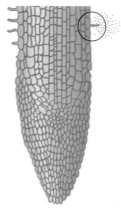

Plant root releases
flavonoid compounds.

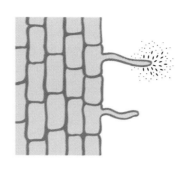

Rhizobium are attracted to root
hair and release carbohydrate.

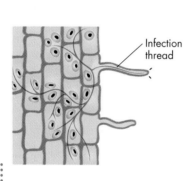

Infection
thread

Infection thread grows through
root, releasing bacterial cells.

Root cells are stimulated
to divide, forming root
nodule.

Rhizobium bacteria infect a plant root by forming a
thin process—called the infection thread—that
penetrates a root hair. Inside the root they multiply to
form the large nodule.

provide food to the host in the form of sugars that
they have photosynthesized, and they also produce
oxygen as a waste product. In some cases, these associations have evolved to the point at which the individual organisms cannot live without their partners.

Living Sands

Foraminifera are amoebalike protozoa that, surprisingly for creatures of their size, build multichambered
shells around themselves, sometimes with hundreds of
chambers. Through pores in the shell walls they extend long, slender "granular" pseudopodia that unite
to form networks in which food particles become
caught. Most foraminifera are barely visible, but unusually large ones may range in size from a millimeter
to several centimeters.

Large foraminifera cells form symbiotic associations with a large number of small algal cells. Known
as "living sands," because the particlelike shells rest on
the sea bottom like grains of sand, these organisms are
often abundant on the floor of shallow tropical seas.
In a square meter of bottom, they can form a kilogram
of calcium carbonate per year. In waters that are particularly low in nutrients, a single organism only 2 mm

The marine amoebae known as foraminifera form hard
body parts such as shells and the spines that radiate
from this *Globigerina*. The spines may function to
reduce the sinking rate of the cell so that it is more
easily suspended by water currents or to make the cell
harder for predators to engulf.

in diameter can contain more chlorophyll than all the algae in a square meter of sea above it. Since chlorophyll is the ultimate source of the food chain, the foraminiferum has at hand potentially much more food, in the form of organic molecules passed on by the algae, than it could get by filtering and eating algae from the water column. The large foraminifera of living sands cannot survive without their companion algae and the light needed for photosynthesis.

Most commonly, the algae that become established in the living sands are small diatoms less than 10 μm in size. They are usually from species that are otherwise uncommon in the environment but capable of growing on their own—if sufficient nutrients are available.

In return for the food supplied by the algae, the foraminifera gather scarce nutrients such as nitrogen and phosphorus from food particles and make them available to the algae. The foraminifera may also provide protection from herbivores and the mobility necessary to track optimal light intensities.

Lichens

Lichens are famous for their ability to grow where few other organisms can. They are the dominant vegetation over about 8 percent of the earth's land surface and are particularly evident on surfaces such as exposed rock that are subjected to alternating wetting

Lichens have the unique ability to grow on the surfaces of bare rock, resisting extreme drying and heating in the sunlight. Their waste products help dissolve minerals in the rock and contribute to the formation of soil.

and drying. Some can function without liquid water, if the air contains sufficient water vapor. Some can fix nitrogen—for example, the lichen *Lobaria oregana* common in the Douglas fir forests of the Pacific Northwest makes a significant nitrogen contribution to the trees it grows on. Although they grow slowly, lichens can live for hundreds of years.

A lichen is a symbiotic association of a fungus and a photosynthetic organism (the photobiont). In over half the 14,000 described species of lichen, the photobiont is a one-celled green alga of the genus *Tre-*

bouxia, and, in about 10 percent of the described species, the photobiont is one of several kinds of cyanobacteria. Some photobionts are free-living species, and some have been found only in lichens. The fungus, which makes up the bulk of the lichen, forms a specialized structure called a thallus, consisting of a loose network of hyphae underneath a dense protective covering. The hyphae offer support to the cells of the photobiont, which exist with the loose hyphae in the same layer. Recent molecular genetic studies confirm that a variety of fungi have independently evolved to form lichen associations. That fungi and algal cells have joined up so many times in evolutionary history suggests that the association is particularly successful—and indeed the fungus species involved are far more likely to be found in a lichen than living alone.

The fungi reproduce by forming fruiting bodies that discharge fungal spores, with a range of about a centimeter; only one lichen that we know of discharges algae along with its spores. The most common lichens also reproduce asexually by dispersing small fragments consisting of hyphae from the fungus surrounding one or more algal cells.

Since the fungus and algae living in association appear much more successful than either living alone, it is natural to assume that each partner contributes to the association and receives benefits in return. However, lichens are difficult to study because of their slow growth and because of the very close contact established between the two organisms; and there is a great deal of uncertainty about the nature of the interaction between the organisms. There is even disagreement about whether the symbiotic association is an example of mutualism (in which both symbionts benefit) or parasitism (in which one exploits the other).

Simon Schwendener in 1869 expressed a view still held by many lichenologists:

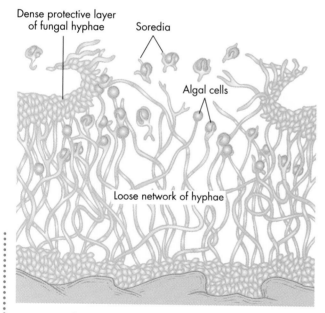

Dense protective layer of fungal hyphae

Soredia

Algal cells

Loose network of hyphae

Most of a lichen's mass is fungus (red), but cells of specific kinds of algae (green) are held in a layer protected from harsh conditions outside the lichen. Some lichens reproduce by releasing "soredia" that contain both an algal cell and bits of fungal hyphae, so that the lichen can be formed immediately on reaching a new location.

The lichens are not simple plants, not individuals in the ordinary sense of the word; they are, rather, colonies, which consist of hundreds of thousands of individuals, of which, however, one alone plays the master, while the rest, forever imprisoned, prepare the nutriment for themselves and their master. This master is a fungus of the class Ascomycetes, a parasite which is accustomed to live upon others' work. Its slaves are green algae, which it has sought out, or indeed caught hold of, and compelled into its service. It surrounds them, as a spider its prey, with a fibrous net of narrow meshes, which is gradually converted into an impenetrable covering, but while the spider sucks its prey and leaves it dead, the fungus incites the algae found in its net to more rapid activity, even to more vigorous increase. (Translation quoted in *The Lichen Symbiosis* by Vernon Ahmedjian.)

To date, research has established the following. As long suspected, the fungus is a clear gainer from the relationship: when isolated from the photobiont in the laboratory, it cannot grow unless supplied with vitamins. In the intact lichen, the photobiont transfers about 90 percent of the carbon it fixes to the fungus, in the form of a sugar. The cyanobacteria in some lichens fix nitrogen, and most of it is transferred to the fungus in the form of ammonia. Although the photobiont seems less obviously dependent on the relationship than the fungus and can usually be grown separately in the laboratory, the fungus probably supplies it with scarce nutrients and also retains moisture and shields it from direct sunlight. In fact, the photobiont on its own would probably be much less successful in the natural environment than as part of a lichen.

External digestion and trading for food are probably the most common ways of obtaining food, but some microbes have developed their own unique ways of getting fed. Many of these microbes attack a living organism using weapons more sophisticated and specific than the release of digestive enzymes. In some cases, the genes of the other organism are manipulated. In other cases, bacteria are used as a weapon.

Culturing Bacteria

Two genera (*Neoaplectana* and *Heterorhabditis*) of parasitic nematodes have developed the remarkable habit of carrying a specific type of bacteria along with them to infect a new host. In a sense, they treat the host as the "soil" in which they cultivate a crop of nutritious bacteria. The bacteria feed on the insect host, and the nematodes feed on the bacteria.

The infective juvenile nematodes do not feed but wait buried in soil or on its surface. A juvenile often extends its whole body away from the soil particles on which it stands, supported only by its tail. When a potential host comes near, it enters the host's body through natural openings (mouth, anus, or spiracles) and penetrates through the thin wall of the gut or trachea. The juvenile then begins to develop into the adult form. As it is developing, it is ingesting some of the host's body fluid, or hemolymph, which causes the bacteria in the nematode's gut to pass out through the anus, "inoculating" the nutrient-rich hemolymph of the insect. The bacteria multiply rapidly, killing the insect within a day or two. Meanwhile, the nematode apparently feeds on the bacteria. So plentiful is this source of nutrients that usually two generations of nematodes and as many as a million progeny are produced within a week or two from a large insect. The last generation, sensing the declining environment, forms infective juveniles, and these disperse from the cadaver, carrying some of the bacteria in their alimentary tract.

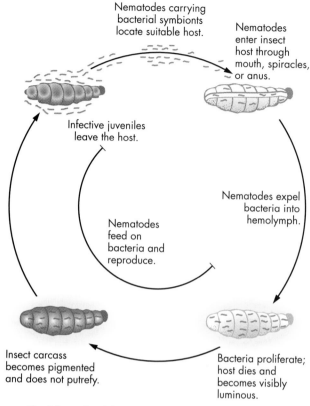

Nematodes carrying bacterial symbionts locate suitable host.

Nematodes enter insect host through mouth, spiracles, or anus.

Infective juveniles leave the host.

Nematodes expel bacteria into hemolymph.

Nematodes feed on bacteria and reproduce.

Insect carcass becomes pigmented and does not putrefy.

Bacteria proliferate; host dies and becomes visibly luminous.

The life cycle of the nematode *Heterorhabditis bacteriophora,* which feeds on a bacterium, *Xenorhabdus luminescens,* that it carries with it to infect insect larvae.

next host. These bacteria are able to keep the insect clear of other bacteria by releasing defective bacterial viruses that destroy other types of bacteria and by excreting antibiotic chemicals that inhibit a broad range of bacteria and yeasts.

The specific bacteria employed in this way have never been found apart from the nematodes that carry them. Apparently, these bacteria have evolved along with the nematodes as this unique mode of parasitism developed. A different species of bacteria is carried by each genus of nematode: *Xenorhabdus nematophilus* bacteria is carried by *Neoaplectana* nematodes, and *Xenorhabdus luminescens* bacteria is carried by *Heterorhabditis* nematodes.

The bacterium *Xenorhabdus luminescens,* as its name suggests, is luminescent, a rare trait among terrestrial bacteria. Indeed, these bacteria give dying insects a faint glow. The function of the glow is

The bacteria *Xenorhabdus luminescens,* carried by the *Heterorhabditis* nematode, are bioluminescent, and the infected insect larvae glow in the dark as seen in this photograph. Bioluminescence is very unusual among terrestrial bacteria, and one can guess that it serves an important function, although what that function may be is not known.

Both the nematode and its bacteria benefit from their relationship. The bacteria are fragile and cannot survive in soil or water, nor can they infect insects on their own. Feeding the isolated bacteria to an insect does the insect no harm, although injecting even one bacterial cell into the hemolymph will kill the insect. Nematodes lacking the bacteria can enter the insect hemolymph but do not reproduce well. Nor do they reproduce as well if other types of bacteria invade the cadaver in place of the customary type. In any case, the juveniles transport only the specific bacteria to the unknown—a good guess might be that it attracts nocturnal scavengers, which eat the cadaver and disperse the nematodes.

Scientists are intrigued by the possibility of using these nematodes for the biological control of insects. The very wide range of hosts they colonize—they infect 10 orders of insects—make their use for this purpose especially promising. The most studied of the nematodes, *Neoaplectana carpocapsae,* infects over 250 insect species. In addition, the infective juveniles can survive for months, even years, when refrigerated.

A Genetic Takeover

The bacterium *Agrobacterium tumefaciens* has evolved an artful strategy for feeding itself. It infects a wounded plant, then manipulates the plant to provide it with food by transferring fragments of its own DNA to the plant chromosomes. The transferred genes direct the plant cell to synthesize and secrete chemicals called opines that the bacteria consume to harvest the nutrients they possess. Scientists have studied this process in much detail because it provides a practical way of introducing foreign DNA into plant cells.

Agrobacterium seems to move up gradients of amino acids and sugars, released by wounded plant cells, until it makes contact with a plant cell and binds to it. There the bacterium is exposed to certain compounds the plant makes to repair the wound, and these activate 25 "virulence" genes. These genes all lie on an isolated piece of DNA, called a plasmid, separate from the bacterial cell's chromosome.

Once activated, some of the plasmid genes prompt the bacterial cell to make a copy of certain other plasmid genes; the copies are then transferred to the plant cell and integrated into a plant chromosome by a mechanism that is not well understood. There the transferred genes become indistinguishable to the

A tiny bacterium, *Agrobacterium tumefaciens,* can have large effects when it infects a plant and causes a crown gall tumor.

plant from any of its other genes: the plant cell carries out their instructions as diligently as if the genes had originally been its own. Some of the genes produce enzymes that synthesize the plant hormone auxin; it stimulates growth and contributes to the formation of the large "crown gall" tumors that protrude from the stems of infected plants. Other genes produce en-

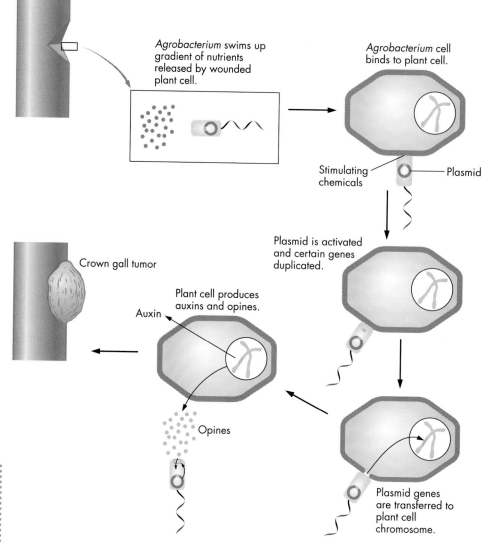

Agrobacterium swims up gradient of nutrients released by wounded plant cell.

Agrobacterium cell binds to plant cell.

Stimulating chemicals

Plasmid

Plasmid is activated and certain genes duplicated.

Crown gall tumor

Plant cell produces auxins and opines.

Auxin

Opines

Plasmid genes are transferred to plant cell chromosome.

The life cycle of the bacterium *Agrobacterium tumefaciens* as it forms a crown gall tumor. This ingenious organism transfers genes to chromosomes of the host plant, and they produce plant growth-stimulating hormones (auxins) and chemicals (opines) that transfer valuable nutrients to the bacteria, allowing them to grow.

zymes that synthesize and transport opines rich in carbon as well as at least one of the valuable nutrients nitrogen or phosphate. The bacterium has in a sense "hoodwinked" the plant into preparing its meals for it.

The opines produced differ from plasmid to plasmid, but they all have some characteristics in common. Whatever the particular opine, the plant cell synthesizes it by linking two different molecules commonly found in cells but not normally linked in this way. Through this method, modest quantities of a single enzyme can produce copious amounts of a novel chemical.

Some of the plasmid's genes stay behind when the rest are transferred to the plant. Now activated by the newly produced opines, they specify enzymes that assist in the breakdown and metabolism of these opines, and only these opines. Thus, each plasmid is a well-integrated system with the capability of producing a certain kind of opine and making use of it. This specificity makes it difficult for unrelated bacteria to make use of opines they have not contributed to. By this complex behavior, then, the bacterium with its plasmid has both provided itself with nutrients and shut out competitors.

Plant-Parasitic Nematodes

Some of the most destructive nematodes that live as parasites on plants have a very specialized life style. A juvenile enters the root of the host plant, establishes a specialized feeding site, and becomes a sedentary sack that remains fixed in place for the life of the nematode. At the feeding site the nematode modifies plant cells surrounding its head to form a suitable food supply. The animal is thought to produce this transformation by injecting, through a hollow, needlelike "stylet," substances that act like plant hormones to modify plant cell development. To feed, the nematode inserts its stylet into one of the cells and sucks up its cytoplasm.

The well-defined glands that these nematodes possess become particularly active immediately after entering a plant, suggesting that they are the source of the hormonelike substances expelled through the stylet. Scientists are still investigating the identity and function of the chemicals involved, but we know these substances act somehow to create feeding sites composed of exceptionally large cells, each with several nuclei. The chemicals must function either by breaking down the walls between cells and causing cell fusion, or by stimulating nuclei to divide repeatedly without the cell also dividing, or by inducing a combination of the two. Specialized ingrowths increase surface area where the giant cells contact vascular tissue and divert plant nutrients from the vascular tissue to the nematode. Nematodes of the same genus produce a distinctive type of feeding site independent of the plant species, so clearly the nematodes control the structure of their feeding sites.

On the average, about 10 percent of all crops are lost to damage inflicted by nematodes. One serious agricultural pest, the southern root-knot nematode *Meloidogyne incognita,* infects more than 2000 species. Where I live in Atlanta, farmers cannot grow ordinary tomatoes year after year in the same location, for these nematodes will accumulate in such high numbers that the tomato plants become stunted and unproductive.

Inducing Cell Breakdown

Microbes also attack mammals, like ourselves, and the difference between those that are harmless to us and those that cause disease is often a behavior that helps the pathogenic type obtain nutrients.

Few microbes grow easily in blood because the iron needed for growth is locked away inside red blood cells; moreover, a protein in the bloodstream tightly binds any iron that escapes. Those microbes that do thrive in our bloodstream are likely to be dangerous pathogens. Many infectious strains produce toxins that break open red blood cells, releasing their iron-containing hemoglobin. A great number of the bacteria that cause disease in ourselves and other animals are known to produce specific proteins that are toxic to the cells in our bodies. In many cases, researchers have demonstrated that strains of bacteria not producing the toxins do not cause serious disease. Thus, it is tempting to speculate that the bacteria produce the toxins because these toxins damage the host cells and thereby make nutrients available to the bacteria.

Consider *Streptococcus pneumoniae,* the common cause of pneumonia. All strains isolated from patients with pneumonia produce the toxic protein pneumolysin. This protein inserts itself into cholesterol-containing membranes and makes them leaky. Since the membranes of animal cells contain cholesterol but those of bacteria do not, the membranes of the bacterial cells remain intact. The toxin is particularly potent in destroying red blood cells, but it has also been found to disrupt the surface of cells lining human respiratory surfaces and to slow the beating of the cilia that sweep debris and bacteria out of the lungs, and it is these effects that lead to pneumonia. Investigators have injected normal strains of the bacterium into the bloodstream of mammals; the bacterial cells multiplied until they attained densities of up to a billion cells per milliliter of blood. The animals died within 24 hours. However, when investigators injected animals with a mutant strain unable to produce the toxin, the bacteria remained at densities a thousand times lower, and the animals showed no detrimental effect even though the bacteria persisted for a week.

Yet, when many cases of toxin production by other bacteria are examined in detail, it is not at all clear how the bacteria benefit from their production of the toxin. In at least some cases, the toxicity is probably an accidental side effect of a chemical that is produced for some other purpose. A clear example is provided by a toxin that causes food poisoning. The bacterium *Clostridium botulinum,* which causes botulism, produces a heat-stable toxin while it grows in food. The bacteria themselves do not normally grow in the body; rather, the victim acquires the disease by ingesting the toxic protein. Consequently, there seems to be little opportunity for the bacteria to benefit from the toxin's effects on the victim.

Bacteria of the genus *Shigella,* which are the source of dysentery in higher primates like ourselves, produce a toxin that inhibits protein synthesis. Consequently, scientists have long suspected that the toxin is an important factor in causing the disease. Volunteers who were fed a mutant strain of *Shigella dysenteriae* that produced little toxin showed milder symptoms of dysentery than those fed a normal strain. Yet genetic mutants that do not produce any toxin still cause lethal dysentery in monkeys, so the toxin cannot be the sole origin of the disease. Another factor is that *Shigella* bacteria have specific proteins on their surface that bind to specific carbohydrates on cells of the colon in such a way that the colon cells engulf the bacteria, and then the bacteria grow inside the colon cells. Ingesting only a few hundred *Shigella* bacteria can lead to serious disease.

The strains of *Corynebacterium diphtheriae* that cause diphtheria carry a bacterial virus that has the genetic information for a toxin that enters animal cells and blocks protein synthesis. Under the right conditions, when iron is scarce, the bacterial cell manufactures the toxin by following the "instructions" on the viral gene. Strains unable to produce the toxin can infect us but do not normally produce disease. It is presumed that the low availability of iron in the mucosal membranes of the host induces the bacteria to synthesize the toxin; it then kills the adjacent cells, releasing nutrients for the bacteria. As the dead cells accumulate, the throat swells and the victim, particularly if an infant, may suffocate.

Whooping cough is caused by the bacterium *Bordetella pertussis*. It produces at least four toxins in addition to proteins that cause mammalian cells to clump or disintegrate. The best known of these toxins is pertussis toxin, which modifies proteins involved in the transduction of signals from the surface of eukaryotic cells. The toxin does not kill the cells but has a more subtle effect on them, and the consequences of its activity are not evident. A serious difficulty in studying this organism is that only humans get whooping cough, so researchers have to perform their experimental studies on animals infected with other, though similar, diseases. In one such study, experimenters altered the bacterium's genes in such a way that the bacterium lost the ability to produce pertussis toxin. These bacteria were less virulent than unaltered cells, so much so that a thousand times more bacterial cells were required to kill a suckling mouse. Evidently the toxin is somehow important to disease in suckling mice, but we cannot be certain that it contributes to whooping cough in humans.

Toxins offer fungi yet another means to catch a nematode. *Pleurotus ostreatus* (commonly known as the oyster mushroom) and some other species in this genus form small secretory branches on hyphae exposed to air. These branches secrete small drops of fluid. When a passing nematode chances to touch these drops, it immediately recoils, and within a few minutes its head shrinks, it stops swallowing, and its whole body becomes immobile. Evidently, the drops contain a powerful toxin. Within a few hours, thin hyphae grow to the mouth of the immobilized nematode and penetrate to the interior, where the hyphae can begin digesting the nematode's nutritious insides.

Many of the examples in this chapter, and earlier ones, have shown that in the micro world, as in the macro world, an organism's survival sometimes comes at the cost of another's life. In such a world, mechanisms of self-defense become of vital importance, and for this purpose, too, the diffusion of chemical agents often plays a crucial role.

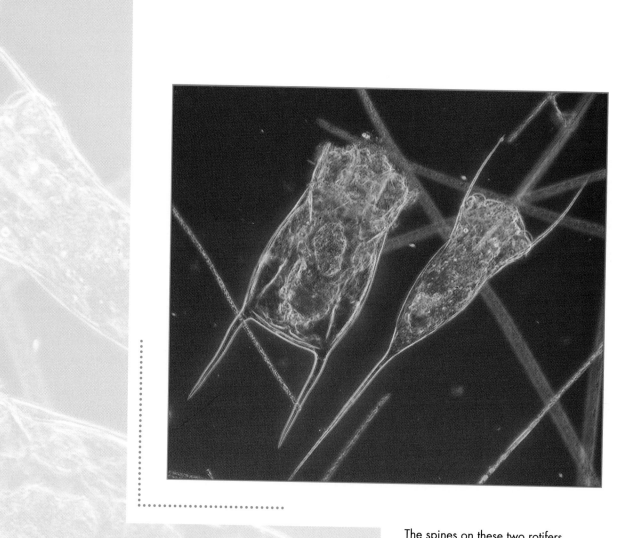

The spines on these two rotifers, *Keratella* on the left and *Kellicottia* on the right, help the animals avoid being eaten by other rotifers and small crustaceans.

Resisting
Others

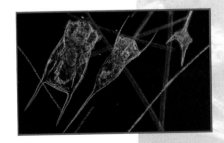

Microbes, like all organisms, are filled with chemicals that are valuable to other life, and so are faced with the fact that there are other organisms around that would gain from eating them. Commonly, protozoa, slime molds, and nematodes eat bacteria. Nematodes and rotifers eat protozoa. And mites, copepods, and other arthropods eat nematodes and rotifers. In addition, almost all organisms now on earth face a competitive world in which many of their neighbors vie with them for nutrients. Microbes that can inhibit, injure, or avoid their competitors and predators are more likely to survive, and most microbes that have been carefully studied exhibit behaviors that help control other organisms in some way.

Most often, microbes use chemicals to poison others. Chemical weapons were probably the first to evolve because the basic mechanisms of life are chemical, and behaviors that developed early in evolution are most likely to be widespread now. Chemical weapons are probably more suitable to microbes because, on the micro scale, currents will not carry chemicals away from the neighborhood where competitors are important. Within the last few decades biologists have discovered that even large plants produce chemicals toxic to animals that might eat them, and many plants also release chemicals that make the surrounding soil toxic to other plants. Poisoning the surrounding environment is particularly appropriate to stationary species, especially those that invest in their local environment by secreting digestive enzymes.

Less frequently, microbes employ mechanical devices such as hard armor or penetrating spines to dis-

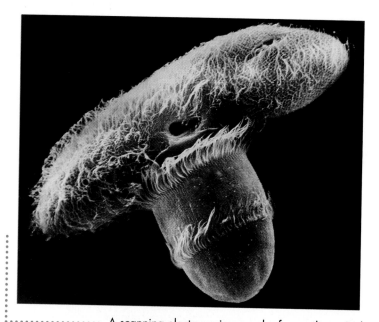 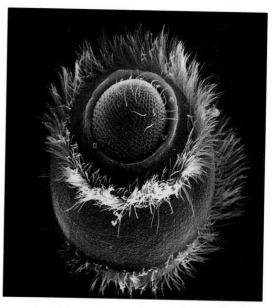

A scanning electron micrograph of one ciliate, *Didinium*, caught in the act of attacking (left) and ingesting (right) another ciliate, *Paramecium*. *Didinium* pokes at every object it touches with a rodlike snout that projects from the center of its front end. Should the snout contact a *Paramecium*, the structure penetrates the cell and *Didinium* swallows the *Paramecium* whole, expanding up to five times its volume to engulf its prey.

courage predators. And in a few cases they appear to send out signals to attract predators of their predators. In looking at some of these defenses among microbes, I will call all organisms that eat another organism predators—even though some scientists would say that an organism that eats algae is a herbivore.

Poisoning in Self-Defense

In lakes and ponds, if the water is warm and nitrogen and phosphate are plentiful, some types of cyanobacteria grow to high concentrations. At the same time, the cells form gas vessicles that buoy them to the surface. Once there, surface currents generated by the winds may carry the cells toward shore, where they accumulate in dense concentrations, forming a colored scum. Animals that drink water containing these high concentrations may grow sick and die. Indeed, entire flocks of birds or herds of animals have died within a few hours of drinking such contaminated water. These mass deaths have become more common where pollution has made nutrients more available to the cyanobacteria.

Such events attract attention, and many have been studied in detail. In most cases, one of three species of cyanobacteria have been implicated. Most strains seem to produce several kinds of toxic chemicals. *Anabaena flos-aquae* produces a neurotoxic alkaloid (anatoxin-a) that paralyzes the muscles, including those of the lung, causing death by respiratory failure. *Microcystis aeruginosa* produces a variety of small peptides that are toxic to the liver of mammals; birds are also sensitive, but cold-blooded animals do not seem to be affected. *Aphanizomenon flos-aquae* produces neurotoxins similar to saxitoxin, a poison also produced by certain dinoflagellates. This is the toxin that causes paralytic shellfish poisoning in humans who eat con-

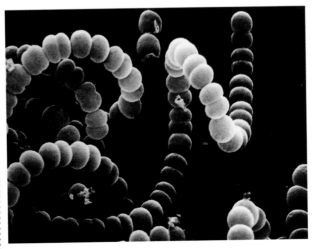

A scanning electron micrograph of *Anabaena flos-aquae* shows the cyanobacterium growing in long chains of cells.

taminated shellfish. The shellfish themselves do not synthesize the toxin but accumulate it in their bodies from feeding on dinoflagellates.

Small planktonic organisms like cyanobacteria are too small to be captured by most large animals but are often caught by "filter-feeding" animals that move water through some sort of screen to concentrate the food before swallowing it. Observers have often noticed that when dense populations of planktonic algae and cyanobacteria are present, populations of filter-feeding zooplankton are low—just when their potential food is most abundant. Why?

As well as being toxic to birds and mammals, cyanobacteria are toxic to filter-feeding zooplankton, such as rotifers and micro-crustaceans, and to many fish. At least some of these animals feed on cyanobacteria, and thus toxic chemicals may help defend the microbes against these predators. Thus, one hypothesis is that the reason that the filter-feeding zooplankton are rare where cyanobacteria are plentiful is that

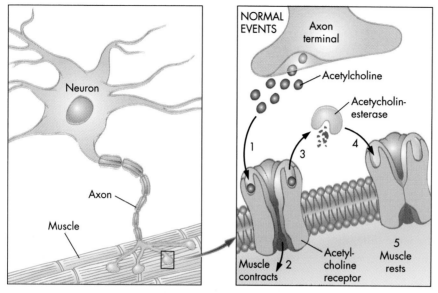
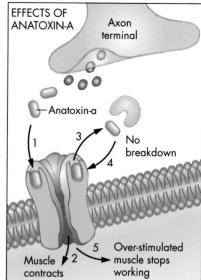

The effect of anatoxin-a on muscle activity. Normal muscle contraction is stimulated by the release of the chemical acetylcholine from the terminal of a nerve cell. After diffusing into a small gap between cells (1 in middle panel), some of the released acetylcholine binds temporarily to a receptor in the membrane of a muscle cell, triggering a series of reactions that result in muscle contraction (2). Within a short time the molecules diffuse back into the gap, where the enzyme acetylcholinesterase splits the acetylcholine in two (3), inactivating it and allowing the muscle to relax (4 and 5). Anatoxin-a binds to the acetylcholine receptor (right panel), preventing the membrane channel from closing, and prolonging muscle contraction. At sufficient levels, the toxin will make all muscles stay contracted, causing paralysis.

the cyanobacteria have poisoned them. In a few cases, it has been demonstrated that the chemicals toxic to the predators of the cyanobacteria are different from the chemicals toxic to mammals. It is hard to imagine of what advantage toxicity to mammals or birds might be to a microbe that lives in aquatic habitats. Perhaps these toxins are also toxic to some predator of the cyanobacteria that has not been tested.

There are, nevertheless, some difficulties with the hypothesis that microbes gain protection from predators by accumulating toxic chemicals. One problem is that the microbe must be ingested and is presumably killed before the toxin has any effect on the predator.

What advantage did a dead microbe gain from synthesizing the toxin?

A possible explanation is that the microbe is protecting not itself, but its closely related neighbors. Microbes often reproduce asexually to produce a clone of genetically identical individuals. If the members of a clone stay close together, so that a predator is likely to prey on more than one member of the clone, a toxic microbe that sacrifices itself saves its neighbors and copies of its own genes. Thus, there could be a selective advantage for clones that synthesize toxins.

Another possibility is that the toxin acts on the predator before the predator kills the microbe. Planktonic microbes are so small and far apart that it had seemed unlikely that they could release enough of a toxin to be effective before being engulfed and killed. However, investigators have observed in some experiments that micro-crustaceans stop feeding after being exposed to just the solution surrounding organisms in a cyanobacterial bloom. If the toxin acted very rapidly, a microbe would only need to protect a small volume around itself. Copepods have been observed rejecting toxic cyanobacteria that they have captured; their fussiness suggests that a toxin need act only on contact to provide the cyanobacteria protection from these predators.

Still unknown is whether predators refuse to feed because they have detected a chemical that is itself toxic or a chemical that functions as a warning signal that toxic microbes are present. In the latter case, the chemical might be a waste product released by the algae that the predator has evolved the ability to detect for its own protection.

Some other features of cyanobacteria may make them an unfavorable source of food for many organisms. Their tendency to grow in the form of long filaments, and to sometimes form large colonies, makes them difficult for many micro-crustaceans to ingest. Some microbiologists have also suggested that cyanobacteria are not easily digested or lack essential nutrients.

A bloom of the dinoflagellate *Gymnodinium splendens,* which occurred in May 1980 off the California coast, also showed subtle signs of toxicity. This bloom migrated between depths of 12 meters, its daytime depth, and 22 meters, its depth at night. The bloom should have been providing micro-crustaceans with a good meal, yet detailed studies revealed that micro-crustacean herbivores were present at lower densities at the depth of the bloom than above or below it. Moreoever, the micro-crustaceans moved in concert with the bloom, as though they were actively avoiding it. Scientists carefully examined the gut contents of herbivores they had removed from the bloom and measured the filtration rates of herbivores still in the bloom. The empty stomachs and low filtration rates showed that the herbivores in the layer of concentrated dinoflagellates were not feeding. The simplest explanation is that the dinoflagellates were toxic.

A bloom of toxic dinoflagellates appears as this "red tide" along the coast of New Zealand.

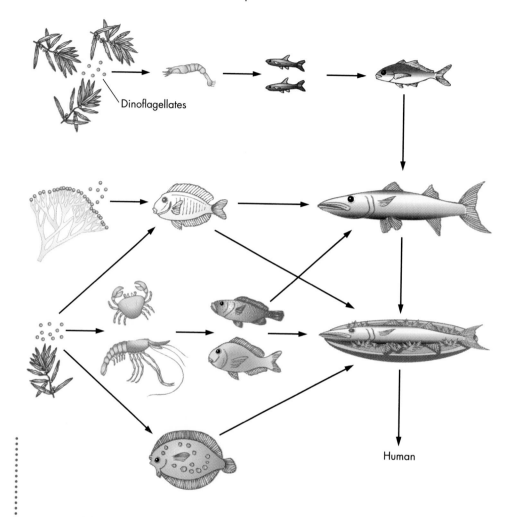

Dinoflagellates

Human

Several food chains that lead to the illness ciguatera: one at the sea bottom (bottom), the classic route (middle), and one at the surface (top). Toxic dinoflagellates living on corals and seaweeds are ingested by grazing crustaceans or fish and remain in their bodies. At the sea bottom, detritus-feeders such as flounder may ingest dead crustaceans or fish with dinoflagellates in their guts, while at the surface herring and anchovies may consume zooplankton that have fed on dinoflagellates. In the classic route, reef-dwelling fish such as groupers obtain the toxin by feeding on crustaceans. The carnivorous fish accumulate the toxins from the herbivores they eat. It is the carnivorous fish that fishermen often catch and serve for dinner.

Certain dinoflagellates are thought to be responsible for the human disease ciguatera (in addition to causing paralytic shellfish poisoning). The inhabitants of tropical coastlines have long known that certain kinds of fish are sometimes toxic to people who eat them, and people often avoid eating fish from certain localities because of their belief that the fish are likely to be harmful. Sometimes the whole crew of a ship becomes sick after eating a large predatory fish, such as a barracuda. Symptoms begin with abnormal skin sensations and may progress to loss of coordination, slowing of the heartbeat, and diarrhea. Most people recover within a few days, but sometimes a victim dies.

About twenty-five years ago, scientists ascribed the cause of ciguatera in the Pacific to toxins produced by the dinoflagellate *Gambierdiscus toxicus,* and since then they have found other dinoflagellate species that also produce the disease-causing toxins. The standard hypothesis is that herbivorous fish ingest the toxic dinoflagellates while grazing on the seaweed on which the dinoflagellates grow. Predaceous fish then eat the herbivores and become toxic to people eating them. Apparently the fish are themselves insensitive to the toxins involved.

Mushrooms are valued for their unique tastes, but feared for the potent toxins that some species produce. A few can even be fatal. *Amanita phalloides* (the "death cap"), *A. virosa* (the "destroying angel"), and several other species produce two types of toxins—phallotoxins and amanitins. Several hours after inges-

Amanita mushrooms, like this *Amanita muscaria,* are notorious for their toxicity. *A. muscaria,* mentioned in folklore around the world, is commonly called fly agaric because it was used to make a fly poison.

tion, liver enzymes convert the phallotoxins into a chemical that attacks the liver cells, leading to extreme pain and vomiting. Amanitin, which blocks protein synthesis, exerts its effects a day later. The widespread damage it causes may last several weeks and lead to permanent liver damage, if not death.

The toxins of other mushrooms affect the mammalian nervous system. Muscarine (from *Amanita muscaria* and other species) excites the parasympathetic nervous system, which lowers blood pressure by slowing the heart and dilating blood vessels. Muscimol and related toxins (from several *Amanita* species including *A. muscaria*) act on the central nervous system and cause hallucinations. Psilocybin and related toxins (notably from the genus *Psilocybe*) are powerful hallucinogens related to LSD. Some historians have even speculated that the Salem witch trials were directed against people who had hallucinated after eating food contaminated with toxic fungi.

The most obvious hypothesis is that the toxins help protect the mushroom's fruiting body from being eaten before it can disperse its spores. The hypothesis has several flaws, however. The long delay between the time the mushroom is ingested and the time the animal experiences many of the toxic effects might make it difficult for animals to learn to avoid eating the mushrooms. The fungus could help primates or birds learn to avoid the toxic mushroom by providing it with a distinctive color, like the warning colors displayed by many poisonous insects and reptiles. Yet poisonous mushrooms are generally not brightly colored, although fungi commonly produce bright carotenoid pigments. Finally, being eaten is not necessarily detrimental to the fungus; an animal is likely to get spores on its body while eating a mushroom, and it will spread them much farther than would otherwise be expected. Thus the purpose of these toxic

chemicals is not at all clear, and they may serve some function other than protecting the mushroom from being eaten.

Forming Armor, Growing Spines

If you surround yourself with a hard material, you make it difficult for a predator to ingest or digest you. Many species of planktonic rotifers secrete a layer of mucus that more or less surrounds the body. The mucus effectively increases the body size, so that small predators cannot ingest the rotifers. It may also possess sticky properties that gum up the mouth parts of predators.

Some rotifers live permanently attached to aquatic plants or other objects. Most of these species form a protective tube around themselves. Some species secrete a clear gelatinous material to form the tube, as planktonic species do. In other species, the rotifer makes special motions that cause particles to become cemented together into pellets. Each new pellet is placed on the lip of the tube, building it up like a brick wall.

A variation on the armor strategy is to grow spines. With relatively little cost in material, a microbe can become much larger—too large for the mouth trying to ingest it. Spines that are stiff and sharp might also injure a predator during ingestion. For whatever reason, many planktonic microbes grow spines.

The large rotifers of the genus *Asplanchna*, common in ponds and lakes around the world, have been extensively studied by John Gilbert and his collaborators at Dartmouth College. *Asplanchna* feed on a variety of smaller planktonic organisms, including ciliates, colonial algae, micro-crustaceans, other rotifers, and even their own species. An *Asplanchna* swims in random directions until the crown of cilia at its ante-

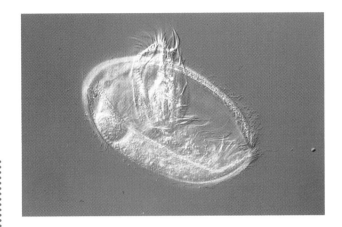

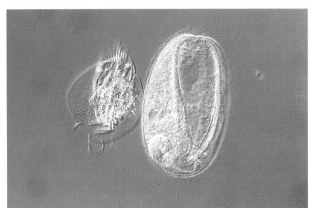

Top left: The large predatory ciliate *Lambadion bullinum* ingests a smaller ciliate, of the species *Euplotes daidaleos.* Top right: After recognizing chemicals released from this predator, a *Euplotes* develops, within several hours, an expanded lateral margin and dorsal ridge, becoming harder to ingest. The *Euplotes*, on the left in this photo, has become too bulky for the neighboring *Lambadion* to digest. Bottom right: In experiments measuring predation rates over a period of one hour, 20 *Lambadion bullinum* captured 23 of the 200 *Euplotes daidaleos* present when the *Euplotes* were smaller than 75 μm in width. But many fewer were captured when the *Euplotes* were larger than 80 μm. Thus only a small increase in size can have a large protective effect.

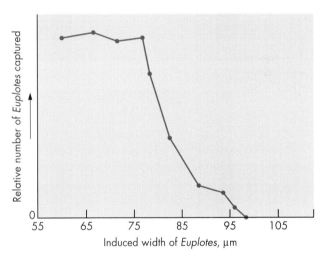

rior end, called its corona, contacts a suitable subject of prey. Thereupon, the rotifer initiates a series of behaviors:

1. It moves its mouth against the prey,

2. the mouth opens,

3. the corona contracts to trap the prey in the pharynx, and

4. the hard, pincerlike jaws manipulate the prey and push it toward the blind stomach.

The rotifer accomplishes the crucial step of recognizing an object as suitable prey at first contact, apparently by evaluating chemical stimuli. Once *Asplanchna* begins an attack, it rarely stops before ingesting its prey. In the laboratory, prey crushed by researchers release chemicals that lead the rotifer to complete its feeding behavior.

Some species of *Asplanchna* are potential cannibals because they capture prey nearly as large as themselves. Many of these species seem to have evolved traits that limit their ability to feed on their own kind. Their large size at birth makes the young harder to capture. They develop extensible outgrowths from the body wall, which make them effectively much larger as prey. And they seem able to recognize and avoid individuals of their own species, probably by detecting chemical stimuli.

Other kinds of rotifers also have defenses to protect against falling prey to *Asplanchna*. These rotifers begin taking defensive action while still in the egg. *Asplanchna* leaks a chemical into the water that alters the embryological development of half a dozen of these prey species. In *Brachionus,* a pair of spines forms that would not otherwise be present, and other spines increase in length. The inducible spines normally point

backward, but when the rotifer is attacked, and retracts its corona in defense, the spines are extended out to the sides. In this position, they increase the effective size of the organism and make it harder to capture. In experiments, *Brachionus calyciflorus* were only one-fifth as likely to be ingested by *Asplanchna* if they had been exposed to *Asplanchna* as eggs.

Since spines provide useful protection, why don't the rotifers always develop them? The answer is presumably that the spines have costs of some kind. Spines certainly require the rotifers to form more cuticle material, and they probably increase hydrodynamic drag. Rotifers that expend energy making spines or moving against higher drag won't have as much energy to put into reproducing themselves. Experiments

Scanning electron micrograph of a cannibalistic giant of the ciliate *Onychodromus quadricornutus* caught in the act of eating a normal-sized member of its species (part of which can be seen on the left). On the right can be seen the four spines that give rise to the species name.

carried out with *Keratella testudo* have demonstrated that individuals with spines do in fact reproduce more slowly than those without them.

Ciliates show even more flexibility in their defenses than rotifers: they can wait until adulthood to grow their spines. *Onychodromus quadicornutus* is a newly discovered large, flat ciliate that moves along the surface in freshwater lakes and ponds. Its exposed upper surface carries a set of four spines, which are unique to this species. When food is abundant, individual cells grow to a length of 400 to 500 μm, but when starved, they become smaller (200 to 300 μm). These ciliates will eat others of their own kind when population densities are high, and a few cannibalistic cells become much larger than normal, as long as 600 to 900 μm. In the presence of the cannibalistic giants, all cells develop longer spines, evidently in response to a soluble chemical that the giants release. The spines also become longer in the presence of the predaceous ciliate *Lambadion magnum*. In experiments, cells with undeveloped spines are far likelier to become another organism's meal than those with well-developed spines. The spines appear to provide protection by making the cells more difficult to engulf.

Escaping

Some swimming microbes are able to move very rapidly for short periods of time, attaining velocities of more than 100 body lengths per second. Although these movements are called "jumps" because of their high speed and short duration, they do not depend on inertia and have little in common with the jumps of large animals. They are really short bouts of high-speed swimming. Some organisms move almost entirely by jumps; in these cases, the discontinuous motion may serve to frustrate small fish and other

predators that hunt by sight. Many other organisms jump only after being disturbed in order to escape closely approaching predators.

The ciliate *Strobilidium velox* cruises at a speed of 150 μm/s, traveling about three of its 45-μm body lengths per second. On occasion, it jumps in a straight line, attaining a speed averaging about 7000 μm/s, equivalent to 150 body lengths per second, for a duration of a second. Such jumps are caused by movements of specialized structures. When undisturbed, *S. velox* spends only about one percent of its time jumping; the distance it travels by jumping, averaged over long times, is about half of the distance it travels by ordinary cruising. This ciliate is occasionally captured by the rotifer *Asplanchna girodi*, but usually when the rotifer contacts the ciliate, the latter jumps and escapes. In one study, *S. velox* escaped 90 times out of 93 contacts with the rotifer. Yet despite the ciliate's skill in eluding rotifers, crustaceans such as the water flea *Daphnia*, which are larger than the rotifer and swim faster, readily capture it.

In Sweden Per Jonsson and Peter Tiselius have studied the feeding behavior of the micro-crustacean *Acartia tonsa*, a copepod that measures about 0.8 mm in length. When the food supply is normal, *Acartia* spends most of its time passively sinking, at 0.6 to 0.8 mm/s. About once a second, it darts upward, taking a tenth of a second for this maneuver, and then sinks again. If a ciliate comes within 0.1 to 0.7 mm of the *Acartia's* first antennae, the copepod turns rapidly (in less than 0.1 second) toward the prey and moves its mouth parts in such a way as to generate a strong current flow toward its mouth. The behavior of the ciliate, which differs among species, has a great effect on whether an attacking copepod snares the ciliate in its current. Copepods capture the ciliates *Strombidium reticulatum* and *Laboea strobila*, which swim at a constant speed of 0.5 to 1 mm/s, more easily than they

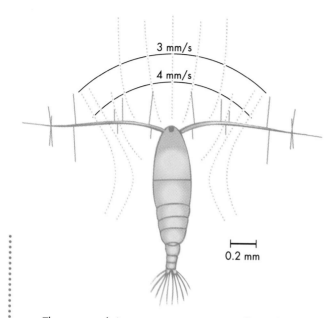

The copepod *Acartia tonsa* generates a flow of water around its body; the current (dashed lines) brings prey closer to the copepod's mouth. Note how the animal's antennae extend out across the path of flow where they can detect the presence of prey in large volumes of water.

the body, but during a jump they independently pivot forward, and then, after all are pointing forward, they independently pivot back to their starting position. Throughout this cycle, the whole animal continues to move forward, suggesting that thrust is generated by forward pivots as well as backward pivots. How the animal keeps moving forward is a mystery, since the paddles appear to be rigid, and the scallop theorem says that the back-and-forth movement of a rigid appendage simply moves a microbe back and forth. Probably, the paddles are not really rigid, and this organism may reach speeds at which inertia has an effect.

The rotifer makes these jumps after making contact with larger rotifers, like *Asplanchna girodi*, or sensing currents generated by feeding *Daphnia*. During careful observations of attempts by *A. girodi* to snatch *Polyarthra*, the larger rotifer captured only 2 of 346 individuals contacting it. *Polyarthra's* escape

capture *Mesadinium rubrum,* which swims in high-speed jumps of less than a second, at a speed of 8 mm/s, and then passively sinks for several seconds.

Using high-speed movie film, John Gilbert has studied the jumps of the small (130 μm) soft-bodied rotifer *Polyarthra vulgaris*. The rotifer cruises at speeds of 300 μm/s (2.6 body lengths per second), but during jumps, the rotifer attains average speeds of 36,000 μm/s (270 body lengths per second). During a single escape response it moves a total distance of about 2 mm (15 body lengths) in 0.06 second. The movement of 12 bladelike appendages, about the same length as the body, generates the jumps. The appendages normally point backward from the sides of

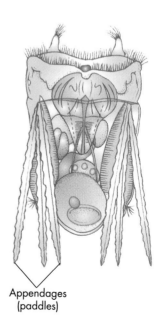

Appendages
(paddles)

The rotifer *Polyarthra* uses its long paddlelike appendages to swim rapidly away from predators.

mechanism is a very efficient one indeed. In other laboratory experiments, *Polyarthra* escaped from *Asplanchna* more than 95 percent of the time, although the small, 1-mm-long crustacean, *Mesocyclops edax,* captured it in about half of all observed encounters.

You may be wondering why the organisms don't always swim at the higher speeds. Presumably the answer is that maintaining these high speeds would demand too much energy. Experts have estimated that while the cost of cruising for some ciliates is about 0.1 percent of the total metabolic energy rate, the cost of jumps is roughly equal to the total metabolic rate during the jump. Clearly it would be impossible to maintain these high speeds for long.

Many small planktonic predators, such as copepods, detect the swimming motions of their prey. If a microbe does not have jumping abilities, it may do best to swim in an inconspicuous way, or stop swimming altogether. The smooth swimming of rotifers may give them an advantage over crustaceans, which swim with a jerky motion. For example, the copepod *Cyclops bicuspidatus* has been observed to respond to water fleas at distances of several millimeters, but it will try to capture a rotifer only when the rotifer is within 0.5 mm and usually only after contact.

When disturbed, many slow-swimming rotifer species retract the delicate corona into the body for protection. This action necessarily brings the animal to a stop, and so probably reduces the mechanical cues its predators use to locate prey.

Wounding the Predator

Toxins may repel and spines injure, but the ultimate in the micro world's aggressive defense mechanisms may belong to certain protozoa that fire projectiles from their surface. We are not yet sure what purpose the

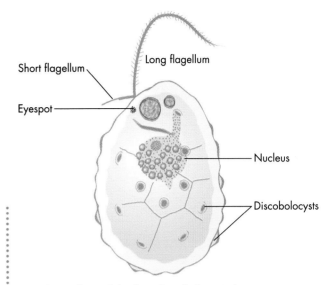

The surface of the flagellated alga *Ochromonas tuberculatus* is covered by 30 to 50 discobolocysts that function as projectiles, presumably to defend against predators.

projectiles serve, but they could well act as bullets directed at predators.

Certain species among the golden algae (Chrysophyceae), for example, have explosive structures called *discobolocysts*. The surface of the biflagellated, yellowish species *Ochromonas tuberculatus* is covered with 30 to 50 of these discobolocysts, which are simple spherical structures, 1.5 μm in diameter, that protrude 0.5 μm above the surface of the cell. When triggered, the top of the sphere is propelled away from the cell, as the mucopolysaccharide contents of the sphere expand into a thread 6 to 11 μm long. The recoil consequently propels the cell about 5 μm in the opposite direction. From the cell's recoil, it has been estimated that the initial speed of the projectile is approximately 100 m/s, or about half the speed of a 0.22-caliber rifle bullet! The projectile might be able to injure a predator that had engulfed the cell, but this remains speculation.

Many ciliates and dinoflagellates have spindle-shaped explosive structures called *trichocysts*. *Paramecium* has about 8000 distributed over its surface. Trichocysts have pointed projectiles on a shaft of protein. Propulsion is generated by a rapid rearrangement in the paracrystalline array of protein molecules in the shaft. During this rearrangement, the shaft increases in length about eightfold.

Attracting Predators of Predators

If subjected to rapidly changing currents, many species of marine dinoflagellates produce brief flashes of bioluminescence. Dense populations of these organisms

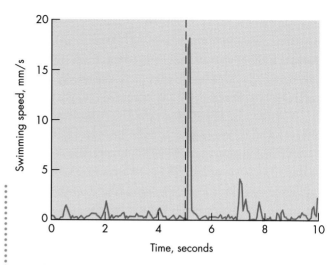

The copepod *Acartia hudsonica* increased its speed more than 10-fold for a fraction of a second after exposure to a flash of light (0.06 second in duration) simulating a bioluminescent flash from a dinoflagellate. The copepod may speed up to get away from the site of the flash, which might attract one of its predators.

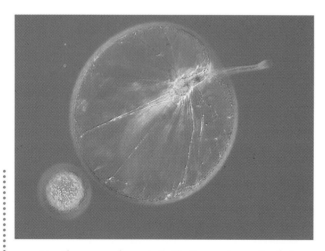

Noctiluca, seen here, is (with *Gonyaulax polyedra*) one of the two dinoflagellates primarily responsible for bioluminescence in coastal waters. It forms a nearly empty buoyant sphere up to 2 mm in diameter, to which is attached a long flagellum used for swimming and a stout tentacle used for catching its prey, often other dinoflagellates.

often provide beautiful displays that attract the attention of people nearby. But what purpose do the flashes serve for the microbe? Their function has been the subject of much speculation but little experimentation. In two experiments, in which scientists placed copepods together with luminous and nonluminous strains of the same species, copepods ate more of the nonluminous strains. This result suggests that the copepods avoid eating luminous dinoflagellates—but why?

One possibility is that the dinoflagellates flash when they sense the feeding currents generated by copepods, and their flashes could draw the attention of fish that might prey on the copepod. If this is the case, the copepod would have evolved to avoid feeding on luminescent prey because of the danger of becoming prey itself. This proposal is called the "burglar alarm" hypothesis.

Several known features of this behavior are consistent with the hypothesis. The color of the light emitted closely matches the color that clear water transmits most effectively and that fish and crustaceans detect most easily. A calculation suggests that dinoflagellates emit sufficient light for a fish to see a meter away.

Many micro-crustacean copepods that could feed on dinoflagellates swim faster and straighter when exposed to artificial light flashes or to a bioluminescent dinoflagellate strain than when exposed to a nonluminescent strain. The light flashes seem to trigger an escape response that puts a distance of several centimeters between the copepod and the dinoflagellate. Overall, several lines of evidence provide good support for this explanation of flashing bioluminescence.

Suppressing Competitors

Most organisms compete with other species for scarce resources such as light or food. If a new variant of a

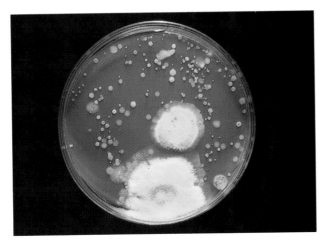

Top left: A variety of microbes have formed colonies on a Petri dish containing agar and nutrients, all from the equivalent of a speck of soil. Sherine Loudermilk, a student at Georgia Institute of Technology, took a soil sample from the banks of the Chattahoochee River, and shook up one gram of the sample in a hundred cubic centimeters of water. She then spread a sample of the liquid one-tenth of a cubic centimeter in volume across the agar of this Petri dish. When she spread other samples from the same solution on plates containing different nutrients, a whole different set of colonies developed. Right: Experimenters grew three species of fungi together or separately, then counted the number of fruiting bodies each formed. In some cultures they also included one or all of four species of bacteria. The area of the rings represents the count of fruiting bodies. Note that the success of each fungus was greatly influenced by other fungi and bacteria in a complex pattern.

Fungi	*Ascobolus crenulatus*	*Chaetomium bostrychodes*	*Sodaria macrospora*	
Fungi separate				No bacteria
				+ Flavobacterium
				+ Methanobacterium
				+ *Pseudomonas*
				+ *Staphylococcus*
The three fungi together				No bacteria
				+ Flavobacterium
				+ Methanobacterium
				+ *Pseudomonas*
				+ *Staphylococcus*
				All four bacteria

species appears with the ability to suppress some of its competitors, it will have an advantage over individuals lacking the ability. Everything else being equal, it will eventually replace them.

Thanks to the enormous developments in the technology of chemical analysis that have taken place during this century, scientists have succeeded in identifying a common mechanism used to suppress com-

Antibiotics and Their Sources

Source species	Antibiotics	Targets of the antibiotic (spectrum of toxicity)
Unicelluar bacteria		
Bacillus polymyxa	Polymyxin-B	Gram-negative bacteria[a]
B. licheniformis	Bacitracin	Gram-positive bacteria[a]
Streptomyces		
Streptomyces nodosus	Amphotericin-B	Fungi
S. halstedii	Carbomycin	Gram-positive bacteria
S. aureofaciens	Chlorotetracycline	Broad spectrum
S. venezuelae	Chloramphenicol	Broad spectrum
S. griseus	Cycloheximide	Pathogenic yeasts
S. erythraeus	Erythromycin	Gram-positive bacteria
S. kanamyceticus	Kanamycin	Gram-positive bacteria
S. antibioticus	Oleandomycin	Staphylococci
S. rimosus	Oxytetracycline	Broad spectrum
S. fradiae	Neomycin-B	Broad spectrum
S. niveus	Novobiocin	Gram-positive bacteria
S. noursei	Nystatin	Fungi
S. griseus	Streptomycin	Gram-negative bacteria
Fungi		
Cephalosporium acremonium	Cephalosporin	Broad spectrum
Aspergillus fumigatis	Fumigillin	Amoebae
Penicillium griseofulvum	Griseofulvin	Fungi
P. chrysogenum	Penicillin	Gram-positive bacteria

[a]Because of a difference in the composition of their cell walls, gram-negative bacteria stain pink to red following the Gram stain; gram-positive bacteria stain blue to purple. Most bacteria are gram-negative, including the spirochetes, *Spirillum*, *Escherichia coli*, and *Shigella*, although *Streptomyces*, *Staphylococcus*, and *Streptococcus* (the causes of "staph" and "strep" infections) are gram-positive.

petitors—the release of toxic chemicals. Many sedentary species, including plants, seem to be in a continuous state of chemical "warfare" with one another, as well as with herbivores. This insight has proved to be of life-and-death importance to our own species: the toxic chemicals developed during the course of evolution yield the antibiotics that often provide an easy cure for once-lethal diseases.

Most of the drugs that we have obtained from these natural toxins act on bacteria. Although protozoa and fungi also cause disease, they are eukaryotes like humans, and there are many fewer molecular differences between them and the human host. A drug that inhibits the pathogen may also do harm to the patient. For this reason, too, it has not been easy to find drugs to fight viral diseases or cancer, since both these types of disease involve nearly normal human cells. In contrast, the molecular biology of prokaryotes differs in many ways from the molecular biology of eukaryotes, and it is relatively easy to find a chemical that targets a molecule existing only in bacterial pathogens.

Soil is normally teeming with busy microbes. A pinch of most any soil, when washed and spread across the nutrient-filled agar in a Petri dish, will yield a thousand colonies of bacteria and fungi, of a wide variety of types. Some of the colonies are surrounded by regions in which other colony types do not grow; experiments have showed that these colonies release chemicals that are toxic to other types of microbe. Most antibiotics have been discovered by looking for the empty zones surrounding a colony in a Petri dish. Today, over 4000 chemicals that inhibit the growth of other microbes have been isolated from bacteria and fungi, and many have become useful as drugs in the fight against bacterial infections.

In most cases, the producing organisms only synthesize the antibiotics when conditions are right— usually when some nutrient (nitrogen, phosphate, or magnesium) other than carbon is in short supply, so that the bacteria halt cell division but continue to synthesize carbon compounds, perhaps for storing energy.

Why does the microbe devote energy to antibiotic production in times of scarcity? One possibility is that at such times the microbe itself, which possesses valuable resources, becomes a more likely object of prey to other microbes. The toxins are the microbe's means of defending itself for a prolonged period until nutrients become more plentiful. Or, the microbe may release antibiotics in order to inhibit nearby organisms that might compete for the scarce nutrients.

If toxic chemicals did function in this way, some scientists expected to find the toxins spread through the environment. Yet researchers have been unable to isolate toxic chemicals from soil and other natural environments, and some microbiologists have suggested that the toxic chemicals isolated from microbes do not function to suppress competitors. Indeed, this issue has become one of the most controversial in the field of microbial ecology. Many microbes synthesize a large number of related molecules that have no clear function, and only some of them are known to act as antibiotics. Some microbiologists have proposed that these chemicals are waste products or food reserves, and that it is only an accident that some are toxic to other organisms.

Recently, biologists have performed experiments in which they place mutated strains of bacteria in natural environments and observe how they do. The strains are identical except that one produces a toxic chemical and another does not. It has been found in most cases that the producing strain is more successful. The experiment suggests that the production of toxins is important and that only technical difficulties prevent us from detecting their presence in the soil.

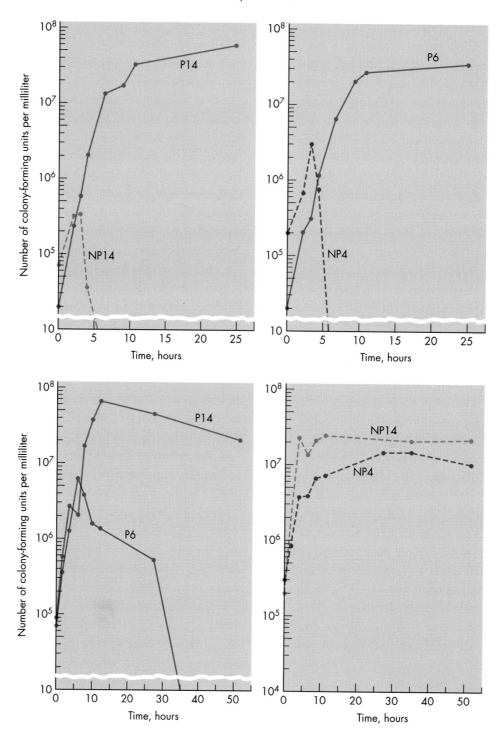

Penicillin

In 1871, John Burdon-Sanderson, a lecturer in botany at St. Mary's Hospital Medical School in London, recorded that a mold of the *Penicillium* group could inhibit the growth of bacteria. Eleven years later, the famous surgeon Joseph Lister cured a chronically infected wound with a *Penicillium* culture at King's College Hospital, London. But sixty more years elapsed before scientists demonstrated the real potential of antibiotics.

The German physician and chemist Paul Ehrlich had been systematically synthesizing new chemicals and testing their ability to kill bacteria without harming the human host. In 1910, the 606th chemical he tested, dioxy-diamino-arsenobenzene (commonly called 606 or Salvarsan), proved to be effective against the spirochetes that caused syphilis, and yet it did not harm the patients to whom it was given. This was the first discovery of a drug useful for treating infections.

Salvarsan soon came to the attention of the British bacteriologist Alexander Fleming, at that time a researcher and teacher at St. Mary's Hospital Medical School. Fleming obtained the first supplies of this drug to be available in London, and his skillful injections of the drug made him much in demand. During World War I, he performed the first comprehensive studies of wound infection and demonstrated the futility of treating wounds with harsh antiseptic chemicals,

which did as much damage to the body's natural defenses as to bacteria.

In November 1921, while stricken with a cold, Fleming made cultures of the mucus from his nose. He noticed that the mucus inhibited the growth of a certain kind of bacteria that had contaminated the culture. Fleming had discovered lysozyme, an enzyme found in many body tissues and secretions that breaks down the cell walls of certain bacteria. The results of further experiments were a disappointment, however. The bacteria in the original culture were especially susceptible, and the enzyme was not very active against most pathogenic bacteria. (Probably susceptible bacteria cannot be pathogenic because the lysozyme naturally occurring in the body easily destroys them.) Another problem was that, when injected into the body, lysozyme persisted for only a few hours. And repeated injections of this protein ran the risk of triggering a fatal allergic reaction. Nonetheless, lysozyme played an important role in the history of molecular biology because it was one of the first enzymes to be crystallized and have its three-dimensional structure determined in atomic detail. Fleming's experience with lysozyme, although disappointing, probably made him better prepared when chance presented him with a truly effective chemical.

After returning to his laboratory from summer vacation in September 1928, Fleming observed that a chance fungus had contaminated a Petri plate on which he had cultured a pathogenic staphylococcus.

Facing page: Experiments have tested the competitive interactions between four strains of marine bacteria isolated from intertidal seaweeds. Two of the strains (P6 and P14) produced antibiotics and two strains (NP4 and NP14) did not. When a producing strain competed with a nonproducer (top two graphs), the producing strain prospered and the nonproducer disappeared. With two producers competing with each other (lower left), one prospered and the other disappeared. But when the two nonproducers competed (lower right), both strains coexisted. This experiment provides evidence that antibiotic production really can suppress other organisms.

He was struck by the fact that, near the fungal colony, the bacteria did not grow. For the next five months, Fleming studied the inhibitory effects of this fungus, using techniques he had developed for his study of lysozyme. He soon found that extracts of the fungus inhibited many pathogenic bacteria, but did not injure the cells of the body's immune system as did antiseptics. The fungus was identified as a *Penicillium,* and Fleming started calling the active mixture "penicillin." Years later it would become apparent that this particular fungus was special—it produced much higher levels of penicillin than most *Penicillium* strains.

Fleming put penicillin to practical use as a selective agent for isolating certain slow-growing pathogens, but its potential as a therapeutic agent did not look promising. Although penicillin was not toxic to animals, it seemed to be inactive in the presence of blood. Furthermore, the penicillin took several hours to kill bacteria but was eliminated from the body in less than an hour. Fleming's published results, which appeared in 1929, attracted little attention. A few scientists obtained cultures of Fleming's *Penicillium* and even treated a few infections with some success, but no one pursued the research.

Meanwhile, Australian-born pathologist Howard Florey had become interested in the function of mucus. His own digestive problems had led him to study lysozyme, and he and his collaborators established that this enzyme split the chemical bonds holding together certain components of bacterial cell walls. After becoming the head of the School of Pathology at Oxford in 1935, Florey hired Boris Chain, a Jewish refugee from Germany who had just received his Ph.D. in biochemistry at Cambridge. Together, in 1938, the two men decided to study penicillin in addition to two other naturally occurring bacterial inhibitors. Interestingly, they expected to answer basic questions about antibacterial mechanisms rather than develop practical therapeutic agents. Within a year, Chain had evidence that the active chemical in penicillin was not a protein like lysozyme but a small molecule. Florey became excited about penicillin's therapeutic potential, and, with generous support from the Rockefeller Foundation, they and other collaborators scaled up the growth of *Penicillium* and the extraction of penicillin.

In May 1940, they had enough material to perform a test on mice. The researchers infected eight mice with a lethal dose of streptococci and injected

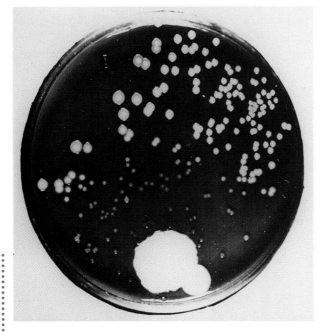

This photograph shows the original Petri dish on which Flemming first observed the inhibition of *Staphylococcus* bacteria by the fungal contaminant *Penicillium notatum.* Note that the bacterium's many round colonies are not all the same size as normally expected—around the large fungus colony at the bottom, the bacterial colonies are much smaller and must be growing more slowly. Later research showed that this fungus secretes the chemical penicillin, which inhibits the growth of bacteria.

•••••• A scanning electron micrograph of the fungus *Penicillium* growing on rotting food.

four of them with a crude preparation of penicillin. Within 16 hours, the untreated mice were dead, but the others survived. Florey and Chain published the results of these and subsequent experiments on mice in August 1940. Treatment of humans still seemed impossible, however: a doctor would have to inject a patient with 3000 times more penicillin to obtain similar concentrations in the human body.

After unsuccessfully trying to interest commercial pharmaceutical companies in producing penicillin on a large scale, Florey took the gamble of turning his Department of Pathology into a production facility, using milk-processing equipment, although he ran a serious risk of fire from the solvents used. In February 1941, his research group treated the first infected patient with penicillin. They injected the patient with 0.1 g

every 3 hours; to obtain enough of the drug they had to extract the penicillin passed in the patient's urine and reinject it. The patient showed dramatic improvement within 24 hours but relapsed after two weeks, when no more penicillin remained available, and died. In August 1941, Florey's group published a paper describing the results of treating six patients. All showed marked improvement, although two had eventually died.

In hopes of expanding the production of penicillin, Florey visited the United States, where he finally convinced a pharmaceutical company to do the necessary large-scale fermentations, which were more typical of the brewery industry than the drug industry. But large-scale production did not start in Britain until Fleming became persuaded of the drug's effectiveness the following year after desperately treating a friend with penicillin that Florey had provided.

Once sufficient penicillin was available (it took about 5 g to treat a person), it rapidly became clear that penicillin was a true miracle drug—it quickly cured many formerly lethal infections. For their discovery of the drug, Fleming, Florey, and Chain shared the Nobel Prize for physiology and medicine in 1944.

In the years since, penicillin and the organisms that produce it have been studied extensively. The chemical is in fact produced by several species of fungi (including *Penicillium chrysogenum, Aspergillus nidulans,* and *Acremonium chrysogenum,* also known as *Cephalosporium acremonium*). Surprisingly, some bac-

teria of the genus *Streptomyces* also produce the chemical, even though it is toxic to most prokaryotes. Apparently, whether an organism is susceptible to penicillin depends on subtle features in addition to the large differences between prokaryotes and eukaryotes.

Penicillin is actually produced in several different molecular forms; each has a common fused-ring structure but varies in the structure of a molecular group attached at one position. All the organisms synthesize it from amino acids via a pathway that also leads to another group of antibiotics, called cephalosporins, which have a different fused-ring component. Both types of antibiotics inhibit the growth of bacterial cell walls and eventually cause the cells to break open.

When scientists cloned and sequenced the genes encoding the enzymes of this pathway, they discovered striking similarities between the genes in the eukaryotic fungi and the prokaryotic *Streptomyces.* The data suggest that these genes were transferred from one group to the other about 0.4 billion years ago, long after the two shared a common ancestor (about two billion years ago).

Antibiotics from *Streptomyces*

Most useful antibiotics are produced by bacteria assigned to a single large genus—the *Streptomyces* bacteria that commonly grow in soil. In fact, the characteristic odor of moist earth is largely the result of volatile chemicals that *Streptomyces* release. These are the same multicellular prokaryotes that we saw earlier could digest the cellulose in rotting vegetation. Although they have a great diversity of growth forms, many grow in long branching filaments that form a mycelium resembling fungi. This filamentous form of growth, so it has been suggested, allows the organisms to bridge gaps between soil particles and makes them too large to be engulfed by ciliate predators. Also like fungi, these bacteria secrete enzymes into the environ-

Molecular structures of various forms of penicillin. Each one has different characteristics that make it more active against certain bacteria and less active against others.

ment that help break down complex organic substrates. All of these properties suggest that they often live in direct competition with fungi and might gain from producing chemicals toxic to fungi.

Fungi are not the only competitors that *Streptomyces* must guard itself against. Because both fungi and *Streptomyces* generally get nutrients by extracellular digestion, they should be vulnerable to faster-growing bacteria that could intercept the nutrients released by their digestive enzymes. Thus, we can imagine that it is particularly important to both *Streptomyces* and fungi to suppress bacterial competitors in their vicinity.

These kinds of observations naturally led microbiologists to hypothesize that *Streptomyces* release antibiotics into the surrounding soil in order to inhibit the growth of nearby microbes that would compete with them for nutrients. However, scientists have not found much direct evidence that antibiotics are synthesized under natural conditions in soil, and some microbiologists believe that they are produced merely as biochemical "accidents."

Nevertheless, a variety of observations support the hypothesis that antibiotics do indeed function to suppress competitors. In one series of laboratory experiments, strains of the bacteria *Escherichia coli* were cultured together with one of several strains of *Streptomyces aureofaciens* differing in their ability to produce the antibiotic tetracycline. When grown in culture alone, in the absence of *Streptomyces*, *E. coli* strains grow two to three times faster than *S. aureofaciens* strains growing alone, so when *E. coli* was cultured together with a strain of *S. aureofaciens* that did not produce antibiotics, the faster-growing *E. coli* replaced the other organism. In contrast, an *S. aureofaciens* strain producing high levels of tetracycline (but with an even slower intrinsic growth rate) dominated the *E. coli* strain until antibiotic-resistant mutants of the *E. coli* appeared, and then the resistant *E. coli* re-

placed the *S. aureofaciens*. After the *S. aureofaciens* was eliminated, the mutant *E. coli* were replaced in turn by *E. coli* sensitive to the antibiotic.

These observations suggest that it is, in fact, possible for microbes to suppress competitors by producing antibiotics. We should not overlook the fact that cells incur significant costs in producing antibiotics and in developing resistance to them. The antibiotic-producing strains of *S. aureofaciens* did not grow as quickly as the strains without this defense, and the speed with which *E. coli* strains lost their resistance to tetracycline once the antibiotic was eliminated also suggests that their resistance was costly. Given these costs, we would not expect to find antibiotic-producing cells or resistant cells in nature, unless there was a continuing need for suppression.

Antibiotics from Other Bacteria

Wheat farmers sometimes face a soil-borne disease called "take-all," caused by the fungus *Gaeumannomyces graminis* var. *tritici*, which attacks the roots of wheat. When wheat is planted repeatedly in the same soil, the severity of the disease sometimes declines with each successive planting. The agent thought to be responsible is any one of several fluorescent strains of *Pseudomonas* bacteria, whose presence in the soil increases the longer the wheat is planted. When added to soil, the strains suppress a variety of soil-borne diseases, including "take-all."

These bacteria produce several derivatives of phenazine, chemicals that act as broad-spectrum antibiotics. Although we may suspect that the phenazine emitted by the bacteria kills the disease-causing fungus, we cannot observe it doing so and must look to more indirect evidence to confirm our suspicions. Here the behavior of specially bred mutant strains is especially illuminating. Mutant strains that do not produce antibiotics also do not inhibit "take-all," nor do

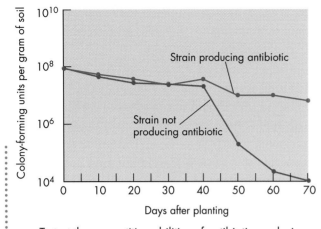

To test the competitive abilities of antibiotic-producing strains in soil, experimenters added to soil a strain of the bacterium *Pseudomonas aureofaciens,* which produces phenazine antibiotics, and a derived strain that has lost this ability. For the first month both strains maintained similar levels, but the nonproducing strain then began a rapid decline. This result suggests that antibiotic production is important for success of these bacteria over long periods.

they survive as long in the soil. Moreover, the antibiotic-producing strains are at a disadvantage in sterilized soil, just as we would expect if the function of the antibiotic was to suppress competitors, but at a cost.

Gramicidins are a family of peptides produced by the bacterium *Bacillus brevis.* Applied to the surface of the skin or eye, they attack certain bacterial infections of these organs. As their mechanism of action is well understood, they provide a good illustration of how an antibiotic can destroy its targets.

Gramicidins are very unusual peptides in that some of the amino acids are in the D configuration rather than the mirror-image L configuration of the amino acids found in proteins. The well-studied gramicidin A contains 15 amino acids alternating in D and L configurations. These molecules insert in bacterial membranes, forming a helix about 3 nanometers long,

about half the thickness of a membrane, with a hollow center. Two such molecules can form a continuous channel through the membrane that is highly permeable to sodium and potassium ions. The ions move through the channel to the other side of the cell membrane, and the cell is forced to waste energy pumping these ions back across the cell membrane to where they were to begin with. In sufficient number, these channels can kill a cell by diverting energy away from required activities.

Nutrient and Vitamin Inhibition

Like ourselves, many microbes cannot synthesize all the kinds of molecules they need and must get so-called vitamins from their environment. These microbes take up vitamins that have leaked out of producing cells into the environment or, if they are predators, they ingest the producing cells. When microbes are dependent on vitamins in the environment, their competitors may suppress them by making the vitamins unavailable.

Certain phytoplankton release carbohydrate-containing proteins that bind vitamin B_{12} and render it unavailable to other phytoplankton that require B_{12} for growth. This mechanism of suppression can be demonstrated in the laboratory. The marine diatom *Skeletonema costatum* grows in culture only when vitamin B_{12} is placed in the culture medium. Even in a vitamin-rich culture medium, though, the diatom's growth can be slowed considerably by adding water in which the dinoflagellate *Gonyaulax tamarensis* had grown. Add high levels of vitamin B_{12}, however, and growth is restored. The dinoflagellate must be eliminating the vitamin, rather than simply emitting a toxin, or adding the additional vitamin wouldn't have overcome the suppression of growth.

Iron is a nutrient required by all organisms to transport electrons within cells. Although iron is an abundant element on earth, most is locked up in highly insoluble iron oxides and hydroxides that form the red clays of highly weathered soils. Consequently, microbes often have a tough time obtaining enough of this element.

Large animals like ourselves exploit this vulnerability by maintaining a low concentration of free iron in the fluids outside cells. In particular, our blood carries a protein (transferrin) that binds iron extremely tightly, reducing the concentration of free iron even more than the formation of oxides and hydroxides does.

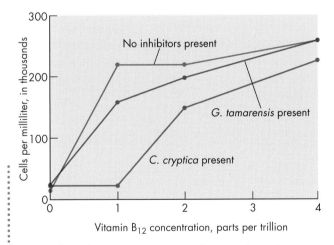

 Without the vitamin B_{12} in its culture medium, the diatom *Skeletonema costatum* does not grow at all, but as little as one part per trillion provides nearly maximal growth after 18 days. Adding solution from cultures of either of the diatoms *Gonyaulax tamarensis* or *Cyclotella cryptica* inhibits growth, but adding more vitamin overcomes the inhibition. *Skeletonema's* recovery indicates that its growth is being inhibited by chemicals released from these organisms that inactivate the vitamin.

Some microbes solve the problem of obtaining iron by excreting an organic molecule, called a siderophore, that binds iron with great affinity. The microbe must of course have some means of retrieving the siderophore once it has bound with iron: specific receptor proteins in the cell membrane recognize the siderophore with bound iron and transport it into the cell.

Only those microbes that produce siderophores can grow in low-iron habitats such as mammalian tissues. Such iron-scavenging systems may be essential to pathogenic organisms that invade host tissues. For example, it appears that only strains of *Escherichia coli* that can produce a siderophore are pathogenic.

By releasing siderophores, one microbe can suppress another microbe and eliminate a competitor that could steal away nutrients. The siderophore binds all soluble iron and makes the iron unavailable to the second microbe, if that microbe lacks the specific receptor proteins for recognizing the particular siderophore structure.

This is thought to be the mechanism by which certain fluorescent *Pseudomonas* species of bacteria suppress their competitors, raising the possibility that adding appropriate *Pseudomonas* strains to soil could eliminate certain bacteria that damage crops. That these bacteria can suppress their competitors is clear, but we are still uncertain that siderophores are the means, although the evidence points in that direction. The first evidence obtained in favor of this hypothesis was that adding iron to the soil removed the beneficial effects of the *Pseudomonas*. Investigators later showed that a *Pseudomonas* strain produces a fluorescent siderophore that when isolated and applied in the laboratory will inhibit pathogenic microbes and stimulate plant growth. In addition, mutant *Pseudomonas* that do not produce the siderophore lose their beneficial effects.

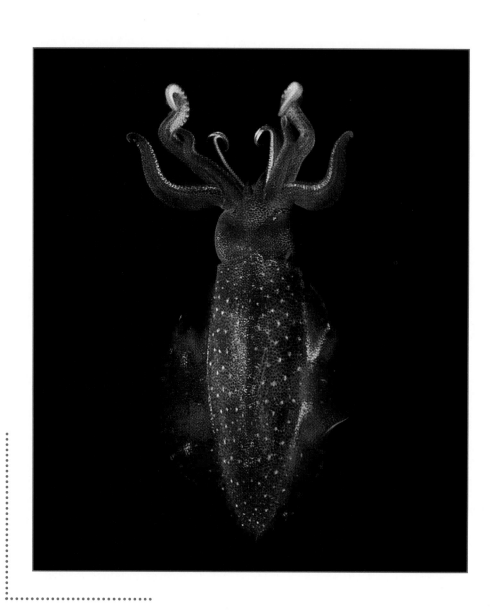

An Atlantic oval squid, *Sepioteuthis sepioidea,* has numerous light organs (light blue spots) scattered across its surface. These mollusks are the most intelligent invertebrates and use bioluminescence to signal to other members of their species. The glow of some bioluminescent squid and fish comes from bioluminescent bacteria harbored in their light organs.

CHAPTER 8

Communicating Without Sound

One of the most surprising biological phenomena is bioluminescence, the production of light by organisms of a variety of kinds. Fireflies emit flashes of light to communicate with the opposite sex, and a few species also lure males of other species with their flashes in order to prey upon them.

At ocean depths where sunlight is dim, fish and other large animals frequently use bioluminescence on their bottom surface to hide the dark silhouette they would otherwise make against downwelling light. In the darkness of the very deep sea, bioluminescence is a common means of communication between large animals. But some microbes, too, are bioluminescent, and for them the function is not so clear. Some scientists in the past have even seriously suggested that bioluminescence is a way for the organism to rid itself of excess energy!

Many bioluminescent fish do not make their own light but harbor specific bioluminescent bacteria in specialized organs. Presumably, the bacteria are paid nutrients in return for their services. The bioluminescent bacteria are mostly flagellated bacilli and are

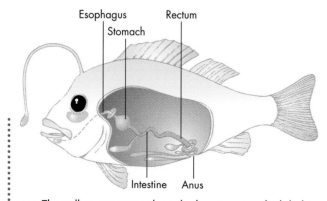

The yellow structures show the locations in which light organs are found in different fish species. In most fish species the organs are associated with the intestine, probably reflecting the evolutionary origins of the light organ bacteria from intestinal bacteria.

rarely found in freshwater or terrestrial habitats. But they can often be found in salt water—not only in the light organs of bioluminescent fish, but also in the intestines of most fish and glowing on decaying animal matter. What could the function of their light be?

One hypothesis is that bacteria evolved the ability to produce light in order to make decaying matter more visible to fish in the darkness of the deep ocean. The fish ingest the decaying material, obtaining food, and in so doing disperse the bacteria. In support of this hypothesis, researchers have found that fish intestines are the most common habitat of bacteria capable of bioluminescence, although the bacteria are not luminescent while in the intestines. In some midwater fish, all colonies that grew from a sample of intestinal contents were capable of bioluminescence. Luminous bacteria not only remain viable during their passage through fish intestines, but they even increase in number.

Most strains are luminescent under some conditions but not others. Except for the nematode symbiont *Xenorhabdus luminscens,* all species are luminescent when the supply of iron is limited and are dark when iron is readily available. Perhaps bacteria growing on detritus are frequently starved of iron but would have access to this element in the intestine of a fish. If this hypothesis is correct, the light emitted by the bacteria on decaying debris is a means of true communication to fish.

Three features of the proposed interaction between the bacteria and the fish suggest that it is a case of true communication. First of all, the light is acting purely as a signal and not doing any work for the fish or "causing" any change to it. Indeed, the fish requires highly specialized organs (its eyes) for the light to have any effect on its behavior. Second, the bacteria emit light for the purpose of signaling to the fish. If the suggestion that bioluminescence is a way of get-

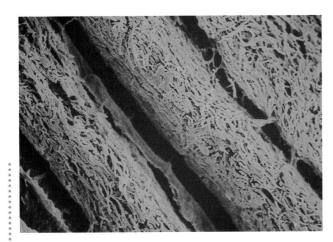

A colorized scanning electron micrograph of the masses of bacteria, *Photoblepharon steinitzi,* found in the light organs of the flashlight fish, *Anomelops.* The light organs of this fish, located right under the eye, are thought to provide illumination for finding food.

ting rid of excess energy were correct, the light signal would simply be an incidental by-product of another activity, and we might say incidental transmission of information was occurring but not true communication. Third, if we had reason to believe that the fish was harmed by ingesting the bacteria, then we might conclude that the bacteria deceived the fish into thinking it had a good piece of food before it. The female fireflies that lure males of other species for prey are engaging in such deceptive communication.

As these examples suggest, my conception of true communication is that it requires an interaction between organisms that is mediated purely by the information in a signal, that the transmitting organism engages in the behavior for the purpose of generating the signal (in more objective terms, it benefits from the interaction), and that the receiving organism also benefits. The signal can be an arbitrarily chosen symbol, although the ease with which the signal can be generated, transmitted, and detected is important. The signal is not required to resemble another signal, as it would in the practice of deception. Thus, the hypothesis suggested above for why free-living bacteria are bioluminescent is an example of true communication.

True communication takes place only between individuals that have a strong interest in cooperation—most often, between potential mates in sexually reproducing species but also between individuals engaging in other cooperative activities or needing to maintain a sufficient separation to avoid competition for nutrients. Most familiar examples of true communication, including insect sex attractants and bird songs, are the attempts of one sex to find a mate of the other sex, but flowers are an example of a plant engaging in true communication with an animal, since their function is to signal to a plant's pollinators that nectar or pollen are to be found. Individuals in the same colony of social insects also have a strong interest in cooperation. All the individuals of a colony are closely related to one another, and the successful reproduction of the queen's genes has much the same consequences as the reproduction of an individual's own genes would. Many examples of true communication between nest mates have been identified in ant and bee colonies. Similar widespread cooperation is necessary among a few microbes when individual cells aggregate in order to form a fruiting body that can more efficiently disperse spores, to secrete enzymes that will degrade a substrate, or to emit light to make a particle visible to fish.

Only a mass of bacteria can produce enough light to be visible to a fish. So, if the hypothesis suggested earlier is correct, and bioluminescent bacteria are trying to attract fish, there should be no reason for an isolated bacterial cell to waste its energy emitting light that no fish could see. And, indeed, in accord with this idea, most species of bioluminescent bacteria emit light only when they are crowded together. How does an individual cell know that it is close to many of its relatives? It assesses the concentration of a chemical signal that they all release into the environment. Such a chemical signal that mediates true communication between individuals of the same species is called a pheromone.

Cells of *Vibrio fischeri,* the best-studied example, start to glow when they are exposed to even low concentrations of the pheromone. Once this critical concentration has been exceeded, a positive-feedback mechanism causes the synthesis of pheromone to increase rapidly. This positive-feedback mechanism works like a switch to ensure that most colonies are either dark or bright, with few in the intermediate state.

Molecular biologists have isolated the genes involved in bioluminescence and have found a gene critical for synthesizing the pheromone located in the cluster of genes needed to produce bioluminescence. This is strong evidence that the pheromone is produced for the purpose of communicating something that will further the aims of bioluminescence. And, thus, there is good reason to conclude that the release of this pheromone is also a case of true communication.

Sound Is Absent

Sound is the stimulus we humans use most for communication, but we will search in vain for an example of a microbe that generates or makes use of sound. Why is sound absent from the world of microbes?

Sound is a mechanical wave created by the back-and-forth movement of molecules in a coordinated fashion. Although the molecules themselves do not travel far, each molecule nudges its neighbors to move, and their combined motion produces alternating high and low pressures and densities that can travel many meters. At typical sound intensities, the average position of a molecule changes only a very small amount. For example, at the threshold of human hearing, the basilar membrane in the ear, responding to a sound wave, vibrates with an amplitude less than the size of an atom.

Animals sense sound by detecting the relative motion of low- and high-inertia structures in their bodies. In the human ear, the movement of the light and flexible basilar membrane relative to the dense and bony spiral lamina (attached to the temporal bone of the skull) stimulates the hair cells, which generate electrical signals sent to the brain. But, as we have seen, viscosity completely dominates inertia in the world of microbes. Consequently, there is no effective way to detect sound.

Another difficulty is Brownian motion. Even if microbes had a motion detector, it would not be useful. All their structures are continually bombarded by the surrounding molecules with sufficient force to cause these structures to move, and they would be unable to distinguish movement due to sound from movement caused by this continuing agitation. In effect, if microbes had a sound detector they would find themselves in a very noisy environment. But none of the sound would carry information; it would all be random "white noise."

Let us take a quantitative look at this argument. How does the speed of molecular motion due to sound compare to that caused by Brownian motion? The average kinetic energy of any particle (one half its mass times its velocity squared) is equal to $\frac{3}{2}$ Boltzman's constant times the absolute temperature. From this it is easy to calculate the corresponding average speed for any size particle. Microbes typically have speeds due to thermal motion in the range 10^{-4} to 10^{-3} cm/s. The speed of molecules due to sound is equal to the square root of the following ratio: the intensity of the sound divided by the product of the density of the medium and the speed of propagation of the sound wave in the medium. For the intensity, consider the following. Animals are able to detect sounds ranging from about 10^{-12} W/cm² (watts per square centimeter) if they are insects or certain kinds of fish, down to 10^{-18} W/cm² if they are cats or dolphins. Within this range of intensities falls the range of ambient sound intensities found in natural environments. If we take the highest of these intensities (10^{-12} W/cm²), the particle speed caused by sound in water is 8×10^{-6} cm/s, which is only 0.01 of the typ-

ical speeds caused by Brownian motion. Clearly it would be difficult to distinguish this motion from the background of noise.

Microbes wouldn't find generating sound any easier. The small size of microbes creates two problems. First, real vibrations are impossible because viscosity dominates inertia, as we have seen, and a great deal of energy would have to be put into driving the motion against viscosity. Second, to generate sound efficiently requires vibrating a structure that is comparable in size to the wavelength of the sound or larger; otherwise

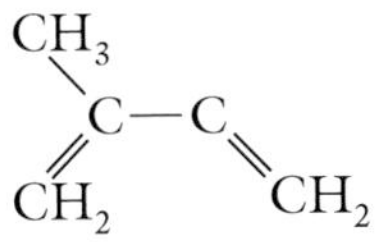

Isoprene

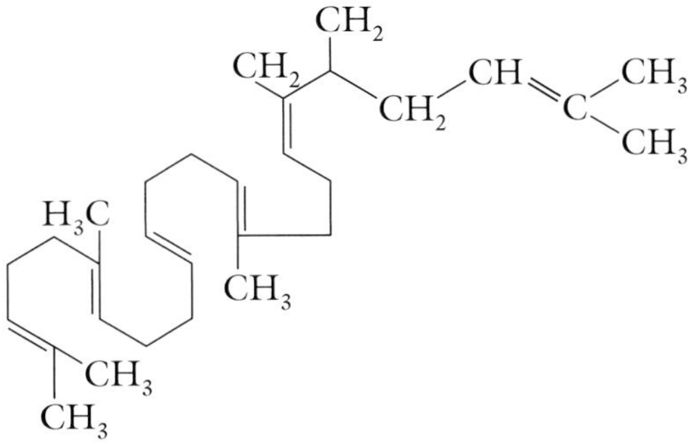

Squalene

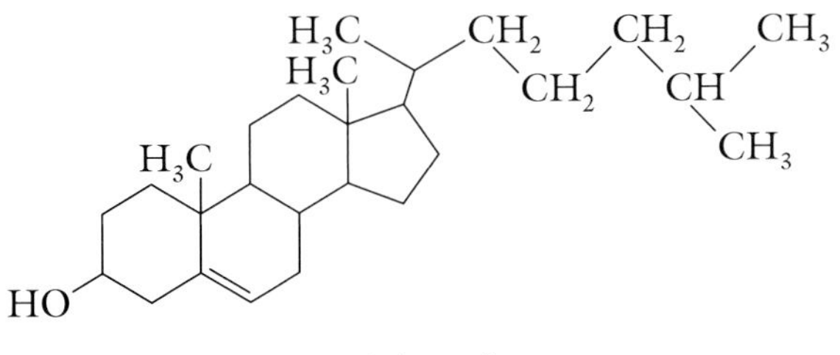

Cholesterol

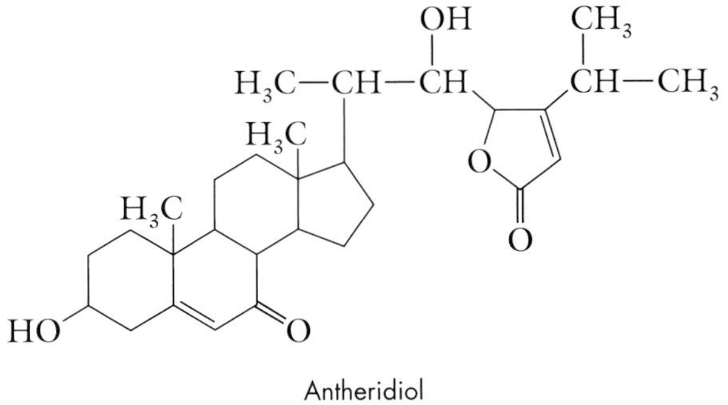

Antheridiol

Organisms use isoprene as a basic building block for forming a wide variety of molecules used as signals, pigments, vitamins, or toxic defenses. In particular, six isoprene units combine in a chain to form the 30-carbon molecule squalene, which undergoes rearrangement to form the 27-carbon molecule cholesterol, the basis of many steroid molecules, including the pheromone antheridiol released by the water mold *Achlya*.

the medium flows around the vibrating object rather than being compressed. An object as small as a vibrating microbe could efficiently generate only sounds of short wavelength. Such sounds would have frequencies in the million-hertz range, which is a thousand times higher than the frequency of ordinary sound. High-frequency sound waves are absorbed so strongly by water that they could travel only a few centimeters even if they could be generated. In short, there are many ways in which the physics of the small-scale world prevents sound from being a useful means of communication.

The only stimuli known to mediate communication between microbes are chemical pheromones. Most identified pheromones fall into one of two general classes—isoprenoids or peptides. Isoprenoids are the most common type of pheromone; these compounds are built on a hydrocarbon skeleton consisting of linked units of the branched, five-carbon hydrocarbon called isoprene. Steroids are a particularly important type of isoprenoid. They are formed by modification of an isoprenoid consisting of a straight chain of six isoprene units (squalene) to form cholesterol, which is the foundation for the synthesis of all steroids. The other major class of pheromones consists of peptides—straight chains of amino acids linked as in proteins. Some pheromones are peptides linked to an isoprenoid. The common use of these two types of molecules can be understood as an easy way for different species to generate a wide variety of chemical structures, using only a few different chemical reactions. For example, there are 20 different amino acids in proteins, and thus there are 20^8, or 26 billion, possible peptides eight amino acids long.

Communicating Population Density

In many microbe species, individuals seem to avoid one another, somewhat like large animals keeping to their own territories. For example, the single-celled amoebae of the slime mold *Dictyostelium* tend to move away from one another and, confined to an area, become dispersed uniformly. Similarly, hyphae of the bread mold *Rhizopus* tend to grow away from their neighbors. In order to keep apart, the microbes must have some way of telling one another where they are. In both cases, individual cells are most likely avoiding chemicals released by other cells of their species. Such behavior could be either true communication, if the chemical is released solely for this purpose, or incidental transmission, if the chemical is a waste product or the release is accidental or incidental to some other activity. Scientists should be able to decide between these two alternatives once they have sufficient information about the chemicals responsible.

Rather than dispersing at high population densities, some microbes improve their chances of obtaining food by adjusting their behavior—like the bioluminescent bacteria. These microbes can also tell when they are in a crowd by sensing chemical signals given off by others of their kind.

For example, only some of the cells in a population of the bacterium *Agrobacterium tumefaciens*, which causes crown gall tumors, carry the infective plasmid, but the carriers transfer their plasmids to other bacterial cells at high cell densities (and in the presence of an appropriate opine chemical released by plant cells in a crown gall tumor). The plant pathogen *Erwinia carotovora*—which digests plant cell walls, killing tissue—produces its digestive enzymes only at high cell density. Similarly, the opportunistic pathogen *Pseudomonas aeruginose,* a bacterium that infects wounds and causes chronic lung infections in immunocompromised people, only secretes its damaging enzymes when its numbers are high enough for extracellular digestion to be cost effective.

Microbes that find themselves living at high densities, and perhaps facing a food shortage, often have

another option—to put all resources into sexual reproduction. It is in assisting sexual reproduction that
pheromones play their richest variety of roles.

Communication for Sexual Reproduction

In ordinary human experience, sexual reproduction is
so common that it might seem it is the only possible
means of reproducing. Although scientists studying
microbes have discovered asexual means of reproduction to be more common, many microbes engage in
sexual reproduction when circumstances are right.

Why do these organisms rely on two different
types of reproduction? In asexual reproduction, an individual simply makes a copy of itself. In sexual reproduction, two individuals combine their genetic

information to produce offspring that are different
from either parent. It is remarkable that sexual reproduction is so common because it has obvious disadvantages. Most clearly, it requires two individuals—if
a single individual reaches a new and favorable habitat,
it cannot reproduce sexually unless another individual
of appropriate sex shows up. This places severe constraints on organisms specialized for colonizing new
habitats, and such organisms frequently have asexual
means of reproduction. Organisms living at low density, far from others of their kind, have difficulties with
sexual reproduction, too. A less obvious disadvantage
of sexual reproduction is that the process of combin-
ing genetic information often provides a mechanism
for the transmission of parasites. This problem is man-
ifest in humans as sexually transmitted diseases.

Given these clear disadvantages, why is sexual reproduction so common—even obligatory in the most

advanced organisms? The most obvious hypothesis is that sexual reproduction speeds up evolution by producing organisms incorporating new combinations of genetic information, received from successful parents. Asexually reproducing species can produce organisms with new characteristics only by random mutation. The obvious success of sexually reproducing species suggests that recombining successful genetic information is much more productive than making random changes.

Many microbes have developed life cycles to have it both ways. They can reproduce either sexually or asexually—depending on circumstances. A common pattern is to reproduce asexually when in a favorable habitat at low population density but, when the habitat begins to deteriorate or the population density reaches a threshold, to begin sexual reproduction, which often produces "resting" eggs or spores that are resistant to harsh conditions and may remain dormant for an extended period of time. These specialized forms are often more easily dispersed to another time or place.

Microbes capable of both types of reproduction have important decisions to make. Should resources be devoted to sexual or asexual reproduction? There are costs associated with either, and the organism that makes the better decision will leave more surviving progeny. Which course is taken is usually decided on the basis of information from the environment—the availability of nutrients or the presence of temperature extremes, acidity, or other unfavorable conditions. In many cases, microbes seem to obtain information about their proximity to other individuals of their species. They may assess the concentration of specific waste products in the environment; alternatively, individuals may release a specific chemical to signal their presence. The aggregate of such signals communicates information about population density.

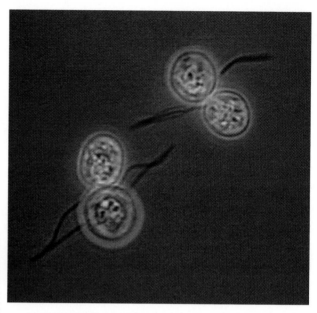

Two pairs of cells of the green alga *Chlamydomonas* mating. Each of these cells has one pair of similar flagella that are longer than the diameter of the cell. The flagella help potential mates find each other by sticking to the flagella of other cells. In this case, the cells themselves are the gametes. Each pair of cells will fuse together, producing a single large cell with four flagella; after a resting phase, it will undergo division to produce four cells that give rise to adults.

There are many mechanisms for bringing the genetic information from two individuals together. In some cases, a whole organism grows or swims toward the other; in other cases, specialized cells called gametes are produced and join together. In some cases, both members of a mating pair behave similarly, but, most often, there is sexual specialization and the members of a pair are of different mating types. Commonly, one mating type (male) is specialized for motility, while the other (female) contributes more cytoplasm to its progeny.

In the three-dimensional world of plankton, where finding members of the opposite sex can be especially challenging, one sex may produce a chemical pheromone to attract the other sex. Terry Snell of the Georgia Institute of Technology and I have calculated the efficiency of various strategies for finding a mate when whole organisms must come in contact with one another. We assumed that one sex (male) was motile but that the other (female) had a choice: she could do nothing to help promote contact; she could swim in random directions; or she could devote the same amount of energy to pheromone production rather than to locomotion. Assuming that the energy available to an organism is proportional to its volume, a straightforward calculation shows that devoting en-ergy to pheromone production rather than to locomotion can make finding a mate thousands of times faster. If the organisms are too small, however, the pheromone diffuses away from the female faster than it is produced, and she is better off using the energy for locomotion. The best estimates we could make from available data suggest that the threshold size at which locomotion and pheromone production are equally effective is somewhere between 0.2 and 3 mm.

Consistent with this calculation, almost all known examples of pheromones that guide an individual of one sex swimming toward the other are produced by animals greater than a millimeter in size. Even rotifers, about the largest of the microbes, are probably too small for pheromone production to be useful. Never-

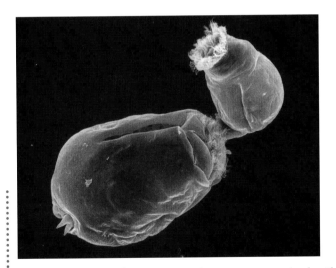 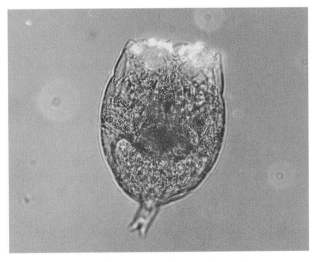

Left: A scanning electron micrograph of rotifers, *Brachionus plicatilis*, mating. The smaller male injects sperm through a penis inserted through any soft part of the unresponsive female. The male locates a female by swimming in random directions until it contacts one, which it recognizes by a specific glycoprotein she carries on her surface. Right: Researchers in Terry Snell's laboratory exposed this female *Brachionus plicatilis* to fluorescent antibodies that bind to the glycoprotein recognized by the male and produce a green glow. They found that the glycoprotein is distributed primarily in the corona of the female, which is where the male usually penetrates her.

theless, planktonic rotifers periodically engage in sexual reproduction (although they commonly reproduce asexually). Both sexes seem to swim in random directions until a male collides with a female. This is an inefficient mechanism, and we would expect that sexual reproduction would be limited to times of high population density. Scientists have calculated the probabilities of rotifers encountering each other; there is a significant chance of fertilization at population densities as low as 1 to 10 individuals per liter, but densities must be as high as 100 to 1000 individuals per liter for mating to approach certainty. And in nature sexual reproduction does indeed seem to take place only at densities higher than one individual per liter.

In contrast to microbes, all groups of crustaceans seem to use chemicals to locate mates, except for the very small "water flea" *Daphnia*. Their use of these chemicals may be one reason that it is so common for organisms above the size limit to have specialized gametes: one sex produces small motile gametes (sperm), and the other sex produces large immobile gametes (eggs) that can release useful amounts of chemical stimuli.

Although pheromones are not an efficient attractant for microbes, they have many other roles to play in reproduction. Consider rotifers again. After colliding with a female, a male needs to recognize that he has contacted a female of his species. Terry Snell and his students have recently demonstrated that in the rotifer *Brachionus plicatilis* the male recognizes a specific glycoprotein on the surface of the female. If the glycoprotein is washed off females and applied to agarose beads, males will attempt to mate with the beads.

In spite of these suggestions that pheromones are of little use to swimming microbes, many other microbes are more or less stationary and have more time to build up pheromone concentrations in their vicinity. Some specific examples will reveal whether they engage in communication.

Plasmid Transfer Between Bacteria

Many bacteria have a primitive form of sex, known as conjugation. One conjugating cell transfers a copy of its DNA to another cell, across a bridge between the cells, although the two cells do not actually reproduce. Thus it might be thought that the cells could use pheromones to facilitate their coming together and forming a bridge. Although we do not know of any examples of pheromones being used in the transfer of the main genome, we do know of a case in which pheromones play a role in the transfer of a plasmid.

Plasmids are relatively small pieces of DNA that contain information for their own replication and sometimes for their transfer to other cells or for a few other functions. The plasmids of many infectious bacteria carry genes that confer resistance to antibiotics or that otherwise influence virulence. One bacterial cell with a plasmid conferring resistance to an antibiotic can transfer a copy of the plasmid to many other cells, and all the cells will then be resistant to treatment with that antibiotic. Even worse, many plasmids combine genes for resistance to several antibiotics and can transmit resistance to all of them simultaneously. This behavior contributes to the spread of drug resistance among bacteria, which is becoming a health problem of increasing concern.

Enterococcus faecalis is a bacterium commonly found in the human intestine, where it does no harm. If it gets outside the intestine, however, it sometimes causes urinary tract infections or very serious infections of the heart. Don Clewell of the University of Michigan has studied plasmid transfer in this species extensively. As is common, a variety of plasmids, containing different assortments of genes, inhabit cells of *E. faecalis*. Some plasmids in this species transfer at relatively high frequency in liquid—about one in a hundred donor cells accomplishes a transfer. In con-

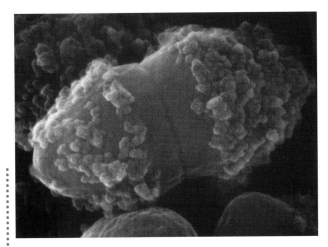

•••••• This scanning electron micrograph shows the bacterium *Enterococcus faecalis*, covered with the fibers that cause it to stick to other cells and so facilitate transfer of a plasmid from one cell to another. The fibers are produced only in the presence of a chemical released by other cells that do not contain the plasmid.

trast, other plasmids transfer so poorly in liquid that fewer than one in a million donor cells will ever transfer one. The more efficient plasmids exploit a pheromone produced by cells lacking the plasmid (recipients). Upon detecting such a pheromone, cells that contain one of the more efficient plasmids (donors) form protein fibers on their surface that cause them to stick to other cells. Cells stick together during random collisions, and, once two cells are attached, the pheromone activates the machinery that carries out plasmid transfer. After a recipient cell has acquired the plasmid, it stops synthesizing the pheromone.

The first such pheromone to be identified was found to be a chain of eight amino acids. At concentrations as low as one molecule per trillion water molecules, this pheromone causes cells to aggregate and increases the frequency of plasmid transfer a thousand-fold. Investigators have identified three pheromones

for other plasmids; all are chains of seven or eight amino acids.

Does the pheromones' signal to the plasmid qualify as true communication? If so, the interaction will benefit the organism on the receiving end of the signal—the donor cell. Yet duplicating and transferring a plasmid costs the cell energy and nutrients, with little obvious return. Here we face a special difficulty in that what is adaptive for the genes in the plasmid may not be adaptive for the donor cell's main genome. In many respects, the plasmid can be viewed as an infectious agent. Taking this view, we could consider the plasmid, not the entire donor cell, as the receiver in the communication, and the interaction is clearly adaptive to the plasmid.

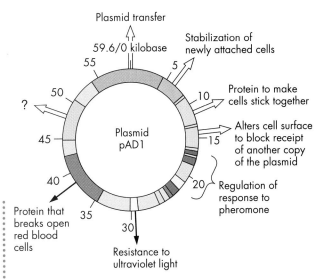

•••••• The arrangement of genes on a plasmid found in the bacterium *Enterococcus faecalis*. The DNA of the plasmid contains 59,600 base pairs and forms a closed loop, with the genes distributed along the DNA. Most of the genes control functions that facilitate the transfer of the plasmid to another cell, and these genes are activated when a pheromone molecule binds to a receptor. The open arrows indicate functions induced by pheromone.

The other question that needs to be addressed is whether the recipient cell benefits from obtaining a plasmid and whether the pheromones' function is primarily to provide a signal for plasmid transfer. The answer isn't clear, but what is known suggests that the pheromone serves other functions. Assuming this to be accurate, the interaction is best classified as incidental transmission rather than true communication. It appears that there are not any clear cases of true communication among bacteria for the purpose of mating, perhaps because they do not engage in true sexual reproduction and thus do not have a strong need for highly specific communication.

Conjugation in Ciliates

Ciliates live most of their lives as individual cells. But under certain conditions two cells sometimes unite for a time and exchange genetic material. The completed exchange leaves both cells with new combinations of the genetic material from the two previous cells. The two new cells then separate and continue their lives as individuals. This form of sex without reproduction, which occurs in bacteria as well as protozoa, is called conjugation.

A cell undergoes conjugation only once. In many species, before even that cell's descendants can engage in another round of conjugation, the cell must go through a number of generations of asexual cell division. In one strain studied, about 30 generations were required. Having satisfied this condition, cells generally become able to undergo conjugation when deprived of food after a period of rapid growth.

Akio Miyake and his collaborators have studied conjugation in the ciliate *Blepharisma intermedium*, with some interesting findings. *Blepharisma* cells are of one of two mating types, and only cells of complementary mating type engage in conjugation with each other. But even two cells of complementary mating type cannot join together to undergo conjugation un-

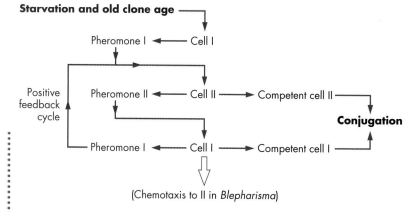

A Pheromone-mediated conversation between two cells initiates conjugation in the ciliate *Blepharisma*. A pheromone produced by cell I stimulates cell II to become competent for conjugation and to produce a different pheromone. This second pheromone stimulates cell I to produce more of its pheromone and to also become competent for conjugation. This mechanism ensures that cells do not get ready for conjugation unless a potential partner is nearby.

less their surfaces have first been altered to make the two cells adhere to one another. That change is initiated by a pheromone released by the complementary mating type. Cells of mating type I secrete pheromone I, which acts on cells of mating type II. After a few hours of exposure, the activated type II cells become "sticky" and secrete pheromone II, which acts on cells of type I to cause them to become "sticky." When any of the "sticky" cells collide, they stick together and proceed with conjugation, if they are of different mating types.

Pheromone I appears to be a moderately large glycoprotein. It is active at the low concentration of 0.1 part per billion, so not much of it needs to be released in order to be effective. Type I cells that are exposed to pheromone II step up their release of pheromone I. The result is a positive feedback loop that causes the production of both pheromones to be enhanced when both types of cell are present.

Pheromone II, in contrast, is a small molecule that begins to show an influence at a threshold concentration of about one part per billion. More recent experiments have demonstrated that pheromone II doesn't just prepare type I cells for conjugation, it acts as an attractant. This is the only known example of a ciliate following the chemical gradient of a pheromone, and the smallest free-swimming organism known to use chemotaxis guided by a pheromone.

Mucorales Mating

In 1924, the German scientist Hans Burgeff performed the first experiment indicating that a fungus releases a diffusable pheromone. Three related species of fungi (*Blakeslea trispora*, *Mucor mucedo*, and *Phycomyces blakesleeanus*) all have two mating types of identical appearance and, as it later turned out, similar mechanisms of sexual communication. Burgeff found

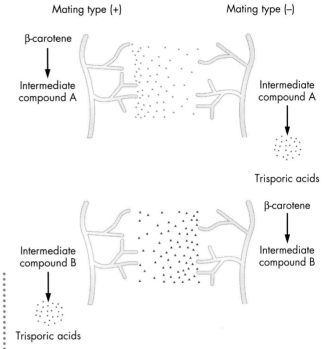

In *Phycomyces* and some related fungi, two hyphae of different mating types recognize that the other type is nearby through their synthesis of trisporic acids. Each mating type forms trisporic acids by a slightly different pathway, but each is missing a step in the pathway and cannot produce trisporic acids on its own. The compound formed before the missing step builds up and diffuses away from the hyphae where it was produced. When two different mating types are close together, however, these intermediate compounds in the synthesis can diffuse to the other type, which completes the synthesis. Thus the buildup of trisporic acids acts as a signal that two different mating types are close together.

that even when the two mating types are separated by a membrane, they initiate the development of the structures characteristic of sexual reproduction. Thirty years later a student of Burgeff demonstrated that separate cultures of each mating type contain chemicals that stimulate the opposite mating type when applied

A water mold growing on a dead salamander. Although water molds are eukaryotes and form hyphae like fungi, their cell walls are made of cellulose, whereas those of fungi are made of chitin. They are important decomposers in their most common habitat, freshwater.

to it. When the two types are grown together in mixed culture, they produce much higher concentrations of trisporic acids than either mating type growing by itself; in the case of *Phycomyces blakesleeanus,* the concentration is at least 100 million times greater.

At first, it was thought that the trisporic acids were the pheromones, but more study has revealed that the pheromones are precursors in the synthesis of the trisporic acids—compounds from which the trisporic acids are synthesized. The best hypothesis appears to be that each mating type is capable of performing different steps in the synthesis of trisporic acids from the isoprenoid β-carotene. Each makes little trisporic acid on its own but releases the product of an intermediate step in the synthesis. The product diffuses to the hyphae of the other mating type, which

completes its transformation to trisporic acid, and much trisporic acid is synthesized. Thus, when the two types are together, much trisporic acid is formed, and this triggers sexual development.

Once both mating types have been appropriately stimulated, both develop specialized lateral branches filled with carotenes, which give them a bright yellow color. These branches appear to grow toward branches of the other mating type in such a way as to meet tip-to-tip. After the tips have come together, they enlarge and produce resting spores.

Water Mold Mating

Water molds of the genus *Achlya* commonly grow on organic debris (twigs, seeds, dead animals) in shallow ponds and lakes, producing a "fuzzy" growth a few millimeters in extent. Like many microbes, they can reproduce either asexually or sexually. Asexually reproducing *Achlya* release flagellated spores, which tend to swim upward because of the difference in density between the front and rear of the cell. These spores are frequently found in high densities near the surface. In contrast, sexually reproducing *Achlya* produce large, thick-walled spores that sink to the bottom, where they may remain dormant for a lengthy period before germinating.

Many *Achlya* species reproduce sexually by self-fertilization, but some species require mating between individuals of two different mating types. Some individuals are capable of playing either the "male" or "female" role, and the choice depends on whether the potential mate is more male or female than itself. The sex roles of two mating individuals is decided by their relative abilities to secrete and respond to pheromones. When hyphae of both mating types grow within a few millimeters of one another, the "male" type develops a long, thin branch, while the "female" type develops a spherical branch. The male branch grows toward the female branch and eventually makes contact with it. Ultimately, their nuclei combine to form several resting spores, each with a new combination of genetic material.

In a series of classic experiments in the late 1930s, John R. Raper demonstrated that at least two complementary pheromones helped coordinate this process. About twenty years later, analytical chemists were finally able to determine the chemical structures of these pheromones: they are steroid chemicals, as are many hormones in the human body. This may seem surprising, since steroids are not very soluble in water and tend to aggregate into clumps. (In the body they are usually bound to other soluble molecules.) Apparently, *Achlya* detect the steroids at concentrations so low that aggregation does not occur.

The female mycelium continuously secretes the pheromone antheridiol, a 29-carbon steroid. When the concentration around a male hypha exceeds 0.01 part per billion, the male produces the long, thin branches, which grow toward the source of the anteridiol. At the same time the male begins producing the other pheromone, oogonial, which induces the female to form the spherical branches. Evidently, these structures release antheridiol at a greater rate

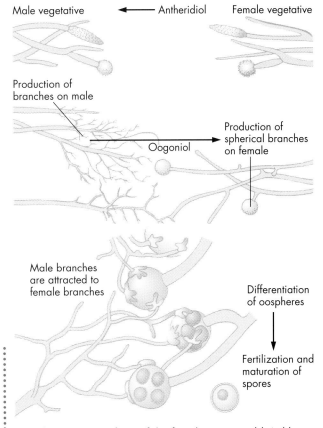

The vegetative form of the female water mold *Achlya* continually produces the pheromone antheridiol. The compound stimulates the male vegetative form to initiate sexual development and to produce another pheromone, oogoniol, which acts back on the female form to induce it to undergo sexual development.

than the unstimulated female hyphae, allowing the male branches to be guided directly to the female branches.

This system of communication appears to be a good solution to the problem of informing the mold of the presence of a potential mating partner in the vicinity. It provides specificity since two chemicals are required, and each can be easily distinguished from other chemicals because of the rigid steroid structure. It is also economical because the high sensitivity requires the release of only small amounts.

Aggregation in the Cellular Slime Mold

When the amoeboid cells of the slime mold *Dictyostelium discoideum* have grown to a high density, they will eventually use up the good food around them. At this point, the cells go into a spore-forming stage. Although they have been living independently of each other, the cells now aggregate to form a relatively large fruiting body that releases spores a few millimeters from the surface. Spores released from this height disperse more effectively, so that their chances of finding a new habitat, with plentiful food, are increased.

In order for the cells to aggregate, each cell must obtain information telling it in which direction to move. This problem has been solved by the development of a unique communication system. By relying on a form of relay signaling using a pheromone, a hundred thousand cells within a radius of a centimeter are able to aggregate to form a migrating "slug," or pseudoplasmodium, which within a few hours forms the fruiting body. The migration of these pseudoplasmodia in soil has already been described in Chapter 5.

In 1967, John Tyler Bonner and his collaborators at Princeton University obtained evidence that the pheromone used by this species was cyclic AMP, which has since been found to be used by animal cells for many internal signaling purposes. Other species of *Dictyostelium* apparently use other chemicals, still to be identified. When a pulse of cyclic AMP from a micropipette is applied to any part of the cell, pseudopodia appear on the side closest to the source.

In addition, each cell briefly secretes cyclic AMP in response to stimulation by another cell's cyclic AMP. Apparently, the binding of cyclic AMP to a specific receptor activates an enzyme that synthesizes new cyclic AMP inside the cell. The cell releases this additional cyclic AMP to the exterior, completing a positive-feedback loop. The amplification created by this positive feedback is soon terminated, however. The cells respond to the increase in the cyclic AMP concentration for only a few minutes before an adaptation process causes them to become unresponsive. Meanwhile, the cyclic AMP outside the cells is being degraded by an enzyme that the cells are continually producing. After the decrease in concentration, cells regain their sensitivity somewhat more slowly.

The mechanism connecting these physiological processes tends to produce oscillations. In a stirred suspension of cells, the cells become synchronized with one another, and the suspension produces pulses of cyclic AMP about every seven minutes. Although this is an unnatural situation, it demonstrates the tendencies of the system to oscillate.

When cells are sufficiently close together on a substrate in a more natural situation, they generate spiral waves of varying cyclic AMP concentration that move outward from centers of activity. The waves have a wavelength of 2 to 3 mm and a period of about 10 minutes; thus they propagate at about 300 μm/min, much faster than the speed of locomotion of the cells

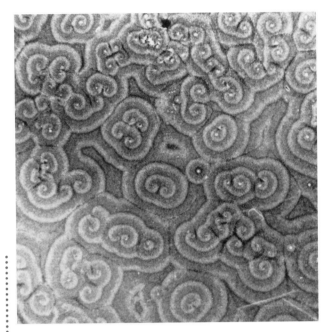 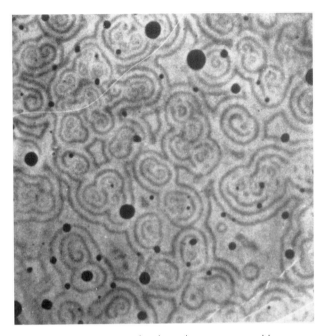

Two ways of visualizing waves of change in cyclic AMP concentration that have been generated by aggregating cells of the slime mold *Dictyostelium*. In darkfield illumination (left), elongated cells that are moving scatter light more strongly and appear brighter than the rounded cells that are not moving. Using a sophisticated technique that they themselves developed, K. J. Tomchik and P. N. Devreotes at the University of Chicago obtained a photograph (right) showing the cyclic AMP concentration directly. The waves, which have a wavelength of about two mm, form spiral or concentric shapes that have been the subject of mathematical analysis.

themselves (about 20 μm/min). Each cell responds by moving in the direction of higher cyclic AMP concentration but only at times when the concentration is increasing. Thus, its locomotion is oriented only about 20 percent of the time. Yet, after about 10 hours, 50 waves have passed, and the cells have moved into aggregates and formed the slug stage.

This is a remarkably sophisticated solution to the problem of aggregation, unlike any observed in other organisms. How does it work? Many fascinated biologists have devoted decades to studying this organism, and we know a great deal about some of the molecular components that generate this behavior, but it is not

yet clear how they are connected to create an overall process.

From these few examples, we can see that microbes have evolved a variety of mechanisms of communication that allow individuals to coordinate their activities. In other situations, microbes could benefit from predicting future events. Although predicting the future is generally impossible, there are some regularities in nature—such as the rotation of the earth—that allow microbes to anticipate certain environmental changes. The next chapter looks at these behaviors.

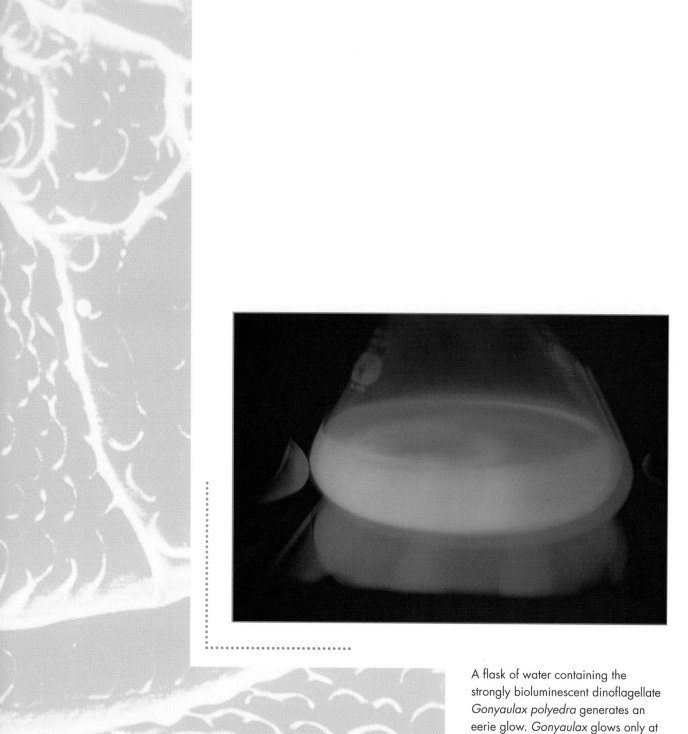

A flask of water containing the
strongly bioluminescent dinoflagellate
Gonyaulax polyedra generates an
eerie glow. *Gonyaulax* glows only at
night, and seems able to anticipate
the fall of darkness.

Anticipating the Future

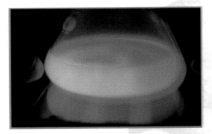

The dinoflagellate *Gonyaulax polyedra*, collected off the coast of California, is one of the dinoflagellates that generates flashes of light in response to mechanical disturbance—probably, as noted earlier, as a defense against predators. In the shallow waters in which the species lives, however, this defense can be effective only at night. During the day, the flashes of light would not be visible to fish against the bright background of scattered sunlight. Since *Gonyaulax* would be wasting energy flashing during the day, it would obviously benefit by restricting its flashing to times of darkness.

Ambient light could inhibit bioluminescence by a simple mechanism. But the organism would then have to maintain high concentrations of the enzymes required for bioluminescence throughout the day, when they were not being used, and these enzymes would be tying up much valuable nitrogen, which is often in short supply. If *G. polyedra* could predict nightfall sufficiently far in advance, it could synthesize the enzymes in time to be ready when they were needed. It could then break down the bioluminescence enzymes in the morning and utilize the nitrogen for photosynthetic proteins during the day. Although we don't know for certain that *G. polyedra* has adopted this strategy, microbiologists have observed this kind of cyclic synthesis and breakdown in luciferase, the main enzyme required for bioluminescence.

Although the strategy appears to be a good one, the catch is the need to predict changes in daylight several hours in advance. Our own species seems to show such an ability—for example, when we wake up just before the alarm goes off. Although it is not possible to predict the future, there are some regularities in nature that allow organisms to anticipate certain changes. Every day the sun appears and disappears—at dawn bringing the energy essential for photosynthesis and the light required for vision, at dusk taking them away. The tides inundate organisms living along the seashore on a fairly predictable schedule. Away from the equator, the yearly cycle of seasons brings important variations in both light and temperature. Such regular periodicities have persisted for billions of years, and many organisms have evolved special mechanisms to take advantage of the regularity to prepare for changes in conditions before they actually occur. These organisms have developed clocklike mechanisms that keep track of time. But can single cells anticipate changes in daylight, for example—or is a brain required?

J. Woodland Hastings of Harvard University has studied this question for over thirty years, and the alga

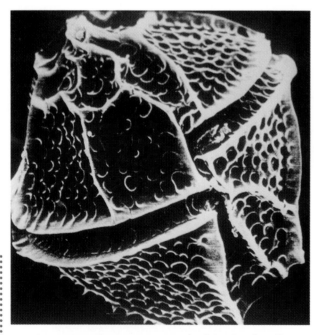

A scanning electron micrograph of the dinoflagellate *Gonyaulax polyedra*. The transverse groove that contains one of its cilia is clearly visible.

G. polyedra has been the favorite organism of choice for his researches, since it is easy to grow, and bioluminescence is easy to measure. Hastings stimulates the cells to flash by bubbling air through the water or mixing the water with acid, and he records their emissions with sensitive light detectors. A cell that has shut down its luminescence function won't flash. One of Hastings's first efforts was to determine whether a clock was cueing *G. polyedra* to turn its luminescence function off and on.

Free-Running Circadian Rhythms

Almost all organisms that live in the presence of sunlight exhibit variations in behavior that follow a daily cycle. Some of these are immediate responses to the level of sunlight, and some are controlled by a clock. How do we distinguish between these possibilities? The answer is to isolate the organism from the periodic variation in light and observe whether the variations continue. When Hastings kept *G. polyedra* under conditions of constant temperature and light intensity, he found that the rhythm of light emission persisted for at least three weeks.

Hastings had not yet proved the existence of a clock, however, for another possibility remained. The microbe could have been responding directly to some unknown stimulus that varied with rotation of the earth. This distinction can be resolved by taking advantage of the fact that a biological clock is never completely accurate. Clocks used for anticipating daily variations, if allowed to run freely without resetting, usually have cycles that are longer or shorter than 24 hours by an hour or so. A clock that is not reset will eventually get out of phase with the daily cycle, while a variation controlled by direct stimulation will stay in

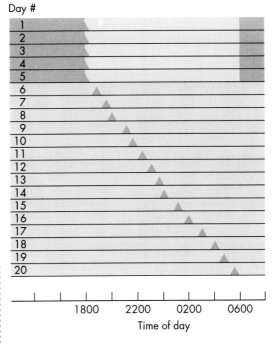

Day #

1800   2200   0200   0600

Time of day

The orange triangles indicate the peak times of bioluminescent glow intensity observed in a *Gonyaulax polyedra* culture. For the first five days, during which the culture was alternately illuminated for a 12-hour period and kept in the dark for a 12-hour period, the glow intensity achieved its peak at apparent "daybreak," though this "daybreak" fell at 1800 hours (6:00 P. M.) local time. When the culture was then subjected to continuous dim light, the time of peak glow intensity steadily drifted with respect to local time, each peak separated from the next by an average of 24 hours and 45 minutes.

phase with the earth's rotation. Consequently, Hastings needed to record the behavior of his *Gonyaulax* for only a few days to demonstrate the presence of a free-running clock. He found that the cycle of bioluminescence exhibited by *G. polyedra* gradually shifted out of phase with the day–night cycle. Their luminescence was indeed controlled by a clock. Such clocks having a period of approximately 24 hours are called "circadian," meaning "about a day."

The biologist who sets out to study biological rhythms must record some behavior over long periods of time, and presenting data becomes a problem. A very useful solution is to break up the time line into successive segments, with each segment equal in length to the 24-hour period. The successive segments are then stacked, with each segment placed below the segment that precedes it. In this arrangement, a vertical line would pass through the same phase of the daily cycle in all segments, and it becomes easy to compare one cycle with another. For example, the middle of the segment might correspond to noon and

the left and right edges to midnight. The activity recorded is often represented by the thickness or density of a horizontal line for each segment, although in the chart on the previous page the activity of flashing is represented by a triangle. In some displays, each line contains two adjacent periods, and each period is presented twice—at the end of one segment and the beginning of the next. This presentation makes it easier to see patterns across the arbitrary boundary between adjacent cycles.

Organisms often exhibit circadian rhythms in a variety of activities. *Gonyaulax polyedra,* for example, exhibits circadian rhythms in cell division, photosynthesis, and a steady weak glow that it maintains in addition to stimulated bioluminescence. An important

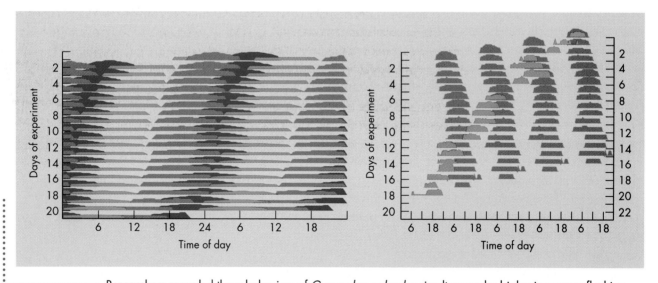

Researchers recorded three behaviors of *Gonyaulax polyedra* simultaneously: bioluminescence flashing (green), steady bioluminescence glow (blue), and aggregation of cells (red). Under constant white light (left), the relative time at which each behavior peaked held constant, suggesting that all three behaviors are controlled by the same clock. Under constant red light (right), however, the time peak flashing and the time of aggregation changed with respect to one another, each now apparently controlled by a clock with a different free-running period.

question is whether each activity has its own clock or whether there is a single master clock that controls all the activities. In most organisms, it appears that most activities are controlled by a single clock. Under free-running conditions, all four circadian activities of *G. polyedra* have the same period and stay in phase with one another, as expected if they are controlled by a master clock. However, it has recently been reported that a rhythm in the swimming behavior of the microbe appears to be controlled by a different clock. Under constant red light, the aggregation of cells has a much shorter free-running period than bioluminescence, and the two types of behavior constantly change in their relative phase relationships.

We have seen that the circadian clock is not particularly accurate; the free-running rhythms frequently have periods that differ from 24 hours by more than an hour. But what about the precision of the clocks? How steadily do the clocks "beat"—that is, how similar are successive periods? How similar are the periods exhibited by different individuals in a population? *Gonyaulax polyedra* cells left under constant red light can continue to glow in synchrony for 19 cycles. Either the clocks in different cells are sufficiently precise to keep individuals in synchrony for many cycles, or the individuals communicate in a way that keeps them entrained to one another.

To test the latter hypothesis, investigators have mixed together cells that were in different phases of the rhythm. Cells that communicate should be able to bring their phases closer together, but no evidence of phase shifting was observed in several species tested. From experiments on the glow rhythm of *G. polyedra*, we know that individual cells can differ in period by about 20 minutes. Thus, the clock has a precision of about $20/(60 \times 24)$, or approximately one percent— a quite impressive degree of precision.

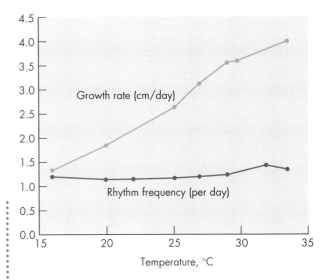

The growth rate of the fungus *Neurospora* (yellow) increases with temperature as expected for most biochemical reactions, but the period of its circadian fruiting rhythm (red) changes hardly at all over this temperature range.

Most biochemical processes speed up two- to threefold when the temperature is increased 10 °C, but the period of free-running circadian rhythms changes very little with temperature. For example, from 16 to 32 °C, the bioluminescence period of *G. polyedra* increases from 22.8 to 25.3 hours, whereas a threefold shortening (to 7 hours) would be predicted from the simplest biochemical models. The growth rate of the fungus *Neurospora crassa* increases 2.4 times between 14 and 30 °C, while the free-running circadian rhythm of spore formation changes little at all. These are but two examples of a general finding that biological clocks have a mechanism of temperature compensation that keeps their period relatively constant in the face of changes in temperature.

Setting Circadian Clocks

Since biological clocks are not very accurate, they would soon get out of phase with the sun's cycle and be useless—if they were not frequently reset. In natural situations, the clocks are continually reset by the daily pattern of change in light intensity, analogous to making an inaccurate watch useful by resetting it every morning. The stimulus that resets the clock is called the "zeitgeber" (German for "time giver") of the clock. For circadian clocks, light is the most common zeitgeber.

In laboratory experiments measuring free-running rhythms under constant conditions, researchers commonly observe that the period of the rhythm is altered by changes in the intensity and color of illumination, if any. This change of clock speed by light is probably the basic mechanism by which the clock is kept in phase with the daily rhythm of the environment. In conditions where the organism is exposed to natural light cycles, the phase of the clock is probably continually shifted over the course of the day. The net shift over 24 hours equals the daily error in the free-running clock.

In some organisms, an increase in light intensity causes a lengthening of the period, while in others the period is shortened. In general, increasing the light intensity causes the free-running clock to run faster in day-active animals and slower in night-active animals. This difference suggests that the setting mechanism may be adjusted to avoid errors at the most critical time of day for both types of animals.

Experiments with lights of different colors indicate that the colors that most effectively set the clock are not the same as those most effective in photosynthesis or in guiding locomotion. This suggests that the clock has its own pigment for detecting the ambient light intensity. The effect of constant light on the bioluminescence rhythm of *G. polyedra* depends on its color—blue light causes a shortening of the period of the free-running clock, while red light lengthens the period. Two different pigments must be involved in detecting the light (although they could be two forms of the same molecule).

Timing Cell Division and Other Circadian Rhythms

It is easy to see that photosynthetic organisms might find it expedient to concentrate their resources on photosynthesis during the day, when light is available, and on other activities during the night. Indeed, it has been demonstrated that, even in constant light, *Gonyaulax* has enhanced photosynthetic activities during the day, and that it and other photosynthetic organisms postpone activities such as cell division until night.

Many kinds of phytoplankton schedule cell division to take place at a set time, most commonly at night. Dinoflagellates in particular are tightly regulated: cell division in *Gonyaulax* and some other dinoflagellates is restricted to a 4-to-6-hour window, usually late at night or early in the morning. Dinoflagellates never multiply faster than one division per day. In contrast, diatoms can multiply much faster, and they have much less tendency to synchronize with the light-dark cycle.

Why cell division so often follows a circadian rhythm is uncertain, since light intensity has no direct role in the process. Scientists have to be cautious in interpreting such rhythms—they may not always be guided by a biological clock. Any complex mechanism such as the biochemical pathways of a cell, including those involved in cell division, will have a tendency to

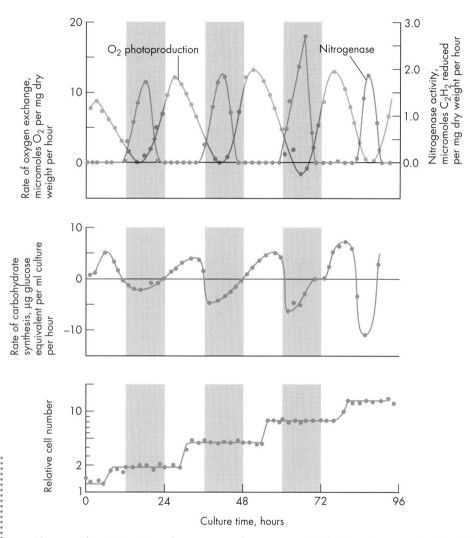

The cyanobacterium *Synechococcus* produces oxygen in darkness (gray vertical bands) and fixes nitrogen, as measured by the activity of one of the enzymes involved, in daylight (top panel). Carbohydrate synthesis (middle panel) and cell division (bottom panel) also show circadian rhythms of activity. On the fourth day, when light was maintained for the whole day, the first three activities continued their rhythms. The experiment was not continued long enough to determine if the cell division rhythm persisted.

oscillate unless it has been specifically designed not to. Certain environmental factors may change the period of the oscillation: heat could speed up the process, or toxic chemicals could slow it down. If these factors are themselves cyclic, they may act to entrain all the individuals in a population to the same phase of oscillation. Thus, it is not surprising that, when cells are in an environment where they divide about every 24 hours, cell division is easily synchronized by daily changes in light, nutrients, toxic chemicals, or temperature. Such rhythms are direct responses to external stimuli rather than to the creation of internal clocks. And we should not jump to the conclusion that such oscillations and zeitgebers are of adaptive advantage to the cell. Nevertheless, there are strong signs that, for many microbes, the timing of cell division is not just a happenstance response to an environmental factor. Many protozoa divide synchronously on a 24-hour rhythm even when successive divisions are several days apart; in this case, only a fraction of the population divides each day. These cycles will even persist with a circadian rhythm under constant condi-

tions, suggesting the presence of a clock shaped by evolution.

It has only recently been discovered that prokaryotes have circadian clocks. Cyanobacteria are confronted by the problem that photosynthesis generates oxygen, and oxygen inactivates their nitrogen-fixing mechanism. Many cyanobacteria alleviate this problem by physically separating nitrogen fixation in special cells, called heterocysts, that are impermeable to oxygen. The unicellular *Synechococcus* has taken a different approach. It confines its nitrogen-fixing activity to nighttime, when photosynthesis is not possible. In this creature, both photosynthesis and nitrogen fixation seem to be regulated by a circadian clock.

The fungus *Neurospora crassa* exhibits a circadian rhythm in the formation of fruiting bodies and asexual spores. As the hyphae of the fungus expand across a surface under constant conditions, the growing front branches more profusely and produces more fruiting structures during one part of the day than during other parts. As a result, the old hyphae and fruiting bodies provide a highly visible record of the rhythm.

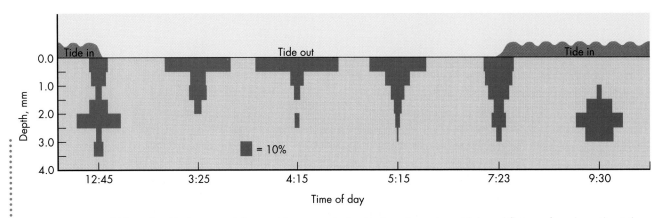

When the tide is out and the mud is exposed, the *Euglena* in an intertidal mud flat are found mostly at the surface, but when the tide covers the mud, they are several millimeters below the surface. The horizontal width of the bars indicates the fraction of cells found at that depth.

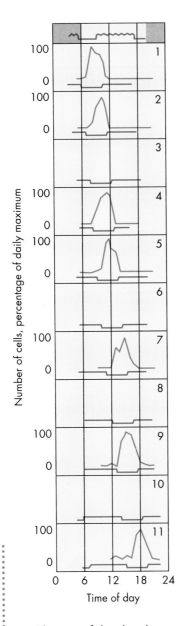
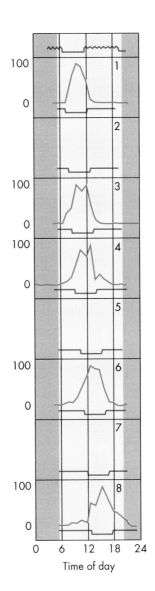

The time of day that diatoms of the species *Hautzschia* migrated to the surface (red curve) kept pace with the expected time of low tide (depression in blue curve) on that day, whether there was a normal light dark cycle (right side) or constant light (left side). These diatoms seem to have a tidal or lunar clock with an accurate period of 24.8 hours.

Although the function of this rhythm is not clear, it has provided a convenient system in which to study the genetics of a circadian rhythm.

Several species of *Paramecium* have circadian rhythms of mating readiness. Tracy Sonnenborn at the University of Indiana and his son David discovered that *Paramecium multimicronucleatum* switched back and forth between two mating types on a circadian schedule. Different stocks of this species switch at different phases of the light–dark cycle and are able to mate with one another at times when they differ in mating type. Individuals thus avoid mating with close relatives and evade the problems of inbreeding.

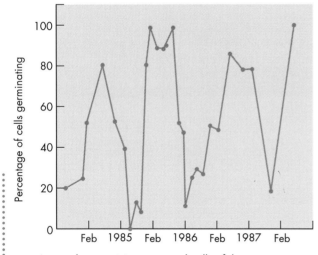

A sample containing encysted cells of the dinoflagellate *Gonyaulax tamarensis* was collected in 1984 for an experiment on germination rhythms. Every month or two a portion was removed from the dark refrigerator in which the sample was kept and placed in conditions suitable for germination. The fraction germinating under these conditions varied throughout the year, but most germination occurred in the spring.

Following the Tides

The tides rule life at the edge of the sea, and an ability to anticipate the changes they bring would be of tremendous advantage. But predicting tides is much more complicated than predicting sunlight. Tides are influenced by both sun and moon and also by the shape of the coastline. Consequently, the tidal rhythm can be complex, and the pattern varies from one coastline to another. The basic period is 12.6 hours, with high tides tending to sweep in when the coast faces toward or away from the moon. Tides are especially strong when the sun and moon are lined up with the earth.

When the tide ebbs during the day, the surface of the exposed sediments along many seacoasts becomes tinted green or golden brown. These colors are the visible sign of vast numbers of microbes that have migrated to the surface in order to obtain the sunlight they need for photosynthesis. When the sunlight is too bright or too dim, or the tide returns, threatening to carry them away, the microbes migrate back down into the sediment. Dinoflagellates, euglenoids, diatoms, and a flatworm containing symbiotic algae have all been observed to behave in this way. Logically, these organisms should find it advantageous to be able to anticipate the returning tide, for then they could migrate back down into the sediment before being washed away. Diatoms and euglenoids do indeed show a tendency to retreat into the sediment before the tide returns.

In laboratory studies, some of these organisms have persisted in their cyclic migrations in the absence of the tide and in constant light. The problem is to decide whether the rhythms are circadian or lunar: the lunar period of 24.8 hours is so close to the solar period of 24.0 hours that investigators find it difficult to determine which period the free-running rhythms are designed to follow.

However, careful studies of the diatom *Hantzschia* from Barnstable Harbor, Cape Cod, have revealed that, under normal light–dark cycles, the timing of emergence drifts across the time of day, keeping in phase with low tide, delayed by 50 minutes each day. After the low tide reaches evening, the diatoms start emerging during the morning low tide, which now occurs during daylight. The diatom seems to possess a lunar clock with a period of 24.8 hours that cues activity at either of two opposite intervals. But it also must have a solar clock, of 24-hour period, that suppresses migration to the surface in the night.

It might be thought that microbes would have little use for clocks with an annual period, and scientists rarely collect data over a span of time long enough to demonstrate rhythms with such long periods.

Nonetheless, there are a few examples. Dinoflagellates sometimes form encysted cells that remain dormant for long periods. Many years later, some may germinate to form vegetative cells that can grow and reproduce rapidly, if conditions are favorable. The germination of a sample of encysted cells of the dinoflagellate *Gonyaulax tamarensis* shows a periodicity under constant conditions in the laboratory that looks very much like an annual rhythm. And it certainly could be advantageous to germinate at a particular season.

Biological clocks are inborn—all *Gonyaulax,* for example, respond to light in the same way from birth to death. An interesting question, addressed in the next chapter, is whether microbes ever change the way they respond to a repeated event. In other words, can microbes learn?

The ciliates in a clump of *Vorticella*, collected in a sewage treatment plant, pump water past themselves and capture food particles from the current. When disturbed, the stalk suddenly contracts, pulling the body away from potential danger.

Learning

The ciliate *Vorticella* usually lives attached to rocks or submerged vegetation. When disturbed by mechanical vibration or a rapid change in water currents, *Vorticella* quickly withdraw into a compact posture. Although it costs feeding time, this contracted position presumably protects the ciliate from predators, which generate water vibrations and currents as they move around. But often *Vorticella* are attached to freshwater animals such as mollusks, tadpoles, or aquatic insect larvae, and this choice of surface creates a problem: the movements of their "hosts" frequently expose the ciliates to harmless vibrations and changes in water currents that could be confused with the approach of a predator. If the ciliates contract, they will lose feeding time unnecessarily. The *Vorticella* found on active animals seem to make some adjustments for their situation—they seem less sensitive to mechanical stimulation than *Vorticella* attached to inert substrates. Have individual *Vorticella* learned that a certain type or degree of mechanical stimulation is harmless?

In experimental tests, investigators have stimulated the ciliates by applying mechanical or electrical shocks. All organisms contracted at the first shock, but fewer responded to succeeding shocks, and 50 percent had stopped responding after 10 to 50 shocks, depending on the stimulus strength of the shock. Even when a ciliate continued responding, the contractions declined in degree and duration as the shocks were repeated. These observations suggest that *Vorticella* may be capable of the simple form of learning called habituation.

We experience habituation when, for example, we stop responding strongly to the repetitions of a loud sound that had made us jump the first time we heard it. If the sound is repeated often enough, we may not respond at all. This behavior can be considered a form of learning in which an organism learns that a certain stimulus that normally triggers a reflex response is no

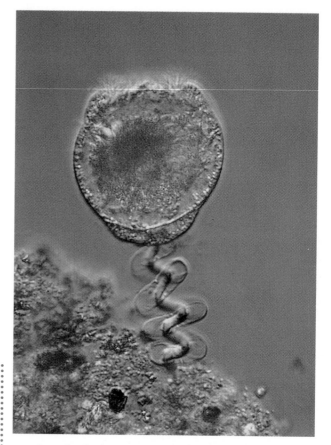

An individual *Vorticella* with its stalk contracted.

longer associated with the reason for the response. We jump in response to a loud noise because such sounds are often associated with danger—we may have experienced pain following a loud sound heard in the past (in which case the response was learned) or our ancestors may have responded in this way and prospered more than those that did not respond (in which case the response is innate). But if the sound occurs in the absence of any danger, the response serves no purpose and can be eliminated.

I have called habituation a form of learning, but what is really meant by "learning"? Basically we mean by the term a process through which experience alters subsequent behavior in a beneficial way. One must be careful in interpreting such a definition, however. Eating may alter the behavior of locomotion by making possible more energetic movements, but this is not learning. By "experience" we mean a pattern of sensory inputs—not events, like an injury or illness, that force new behavior. *Vorticella's* waning responsiveness to shocks appears beneficial in that it allows the organisms more feeding time. But are the shocks acting as sensory stimuli to which the ciliates learn to modify their response? Or are they simply causing injury or fatigue?

To test whether true learning was taking place, investigators let the organisms become habituated to either mechanical or electrical shock, but not both, and then exposed them to the other type of shock. The *Vorticella* responded more strongly to the novel stimulus than the one previously experienced, demonstrating that the response had not waned because of injury or fatigue. Furthermore, the experiment indicates that the learning mechanism distinguishes between these two types of stimuli, and this tells us something about how *Vorticella* perceive their world. Thus an important question about learning is how many different sensory stimuli the organism can learn from without confusing one stimulus with another. The question cannot be answered with any certainty for microbes, because in their studies of these organisms, researchers have rarely used more than the minimal two stimuli necessary to demonstrate a degree of stimulus specificity and eliminate fatigue or injury as explanations.

Another important question about learning is how long the effects last. In other words, how long can memory hold the information learned? The persistence of learning has not been measured carefully in protozoa, but once habituated to a stimulus they have been observed to remain unresponsive to it for less

than a minute in some cases and up to a few hours in others.

A logical consequence of our definition of learning is that an organism learns the time of day when a circadian rhythm is set to the proper phase. Since the phase of the rhythm persists for several days under constant conditions, microbes have a memory for such information that lasts at least this long. But the setting of an internal clock is a very specialized type of learning, and many biologists would not con-

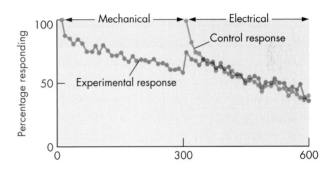

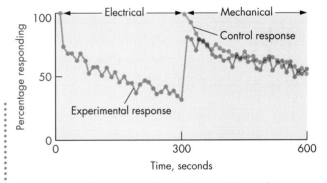

Testing habituation in *Vorticella convallaria*. The fraction of organisms responding to mechanical or electrical shocks, given every 10 seconds, declined rapidly with the first few shocks and then more slowly. After five minutes, the type of shock was switched, and more organisms responded, demonstrating that the response did not wane due to fatigue.

sider it learning at all. They are interested in learning that is governed by general mechanisms and that allows an organism to learn a variety of patterns of information.

In several respects, habituation resembles sensory adaptation or physiological acclimation. We can consider the regularly repeated shocks as a steady stimulus whose intensity is measured by the shock frequency; the waning of the response is like adaptation to a new level of stimulation. It is often found, however, that exposing an organism to another stimulus, especially if it is strong, causes the organism to regain its response to the first stimulus. This kind of undoing of the behavioral change is not observed in sensory adaptation or physiological acclimation, but has often been observed in animals at our own size scale. It can be understood as refocusing of "attention" after receiving new information that suggests a potential threat, and this characteristic is now considered a defining aspect of real habituation. No one has yet demonstrated that protozoa can regain responsiveness to the original stimulus, as would be required of true habituation. If we see this ability anywhere in the micro world, it is likely to be among the tiny animals that have true nervous systems.

Do Nematodes Learn?

The nematode *Caenorhabditis elegans* has become a popular subject of experimental study, especially study of the genetic control of development. Of the hundreds of people doing research with this nematode, a few have investigated the question of learning.

In the laboratory, this organism is commonly grown on Petri plates with an edible lawn of *E. coli* bacteria. As the nematodes crawl around in the bacteria, they may happen to come to the edge of the lawn of bacteria, beyond which bare agar extends. At this

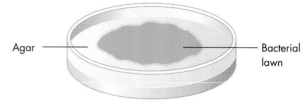

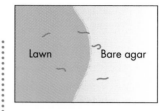

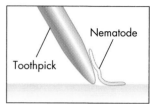

To investigate the escape response of the nematode *Caenorhabditis elegans*, I transferred individual nematodes on a toothpick to a fresh lawn of *E. coli* in a Petri dish. If in their meanderings they came to the edge of the lawn, I recorded the elapsed time and noted whether they went on across the boundary onto the bare agar or turned around to return to the bacteria.

point, they usually stop, turn around, and crawl back into the lawn. In 1971, I noticed that immediately after being transferred to the Petri dish on a toothpick, the nematodes did not make this response—rather they continued onto the bare agar and away from their food. I interpreted this behavior as an escape response triggered by the mechanical stimulation of being handled on the toothpick. Nematodes are preyed upon by other nematodes and mites, and an escape response triggered by mechanical stimulation makes sense.

I wondered how long this altered behavior might last and if it might represent a form of learning. Consequently, I made some careful records of the time elapsed between a nematode's transfer onto the Petri dish and its encounter with the border of the lawn and noted whether the nematode turned back into the bacteria or continued across onto the bare agar. Of 55 nematodes studied, only 3 behaved ambiguously at the boundary. Of 35 nematodes that encountered the boundary within 40 seconds of handling, only one turned around, but 15 of the 17 encountering it later turned back into the bacteria. The lowered probability of turning around at the boundary had a rather well-defined duration of 40 seconds after handling. Clearly the nematodes had some sort of memory persisting this long.

Another feature of nematode behavior that resembles learning is sensory adaptation. Earlier we saw evidence that sensory adaptation to chemical stimuli lasts for a few minutes. More impressive is the length of time for which a nematode remains adapted to some temperature. At a particular ambient temperature, both a nematode's activity and its direction of migra-

tion in a temperature gradient depend on its previous temperature experience. After a large shift in ambient temperature, the first temperature continues to influence behavior for several hours. But again this is presumably a specific physiological response, and most biologists would not consider it a form of learning.

More recently, Catherine Rankin of the University of British Columbia has demonstrated habituation in *C. elegans*. Rankin placed the nematodes in a Petri dish on an agar surface that could be made to vibrate by tapping the dish mechanically. When the surface vibrated, the worms stopped forward locomotion and backed up a short distance. Both the fraction of worms showing this response and the average distance they moved backward declined with successive vibrations, whether the vibrations were spaced 10 seconds

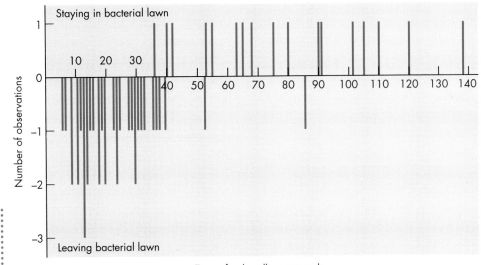

Nematodes that encountered the boundary of the lawn within 40 seconds of getting off the toothpick almost invariably crossed over onto the bare agar. In contrast, those that encountered the boundary later almost always turned around and moved back into the bacteria. I concluded that being picked up on the toothpick had triggered an escape response.

apart or 60. After being subjected to 30 such vibrations, the nematodes could remain unstimulated for at least 30 minutes and still show weaker responses when tested—suggesting a memory of at least 30 minutes. An electric shock caused them to regain some responsiveness to the mechanical taps, although the response never approached the initial level. The experiment suggests that true habituation is occurring, but does not clearly rule out fatigue or sensory adaptation. What about more sophisticated types of learning?

Associative Learning

In what most biologists and psychologists would call real learning, an organism learns an association between a particular sensory stimulus and an important event in the environment—such as between a scent and food. A stimulus, the scent, that is associated with a cause of the event, the food, will be a reliable predictor of that event. Cause and effect are often closely related in time, with the cause necessarily preceding its effect, and stimuli connected with the causecan function as reliable predictors of the effect. Higher animals have built-in mechanisms for learning from such close coincidences. For example, if you flip a light switch and a second later you hear a bang, you are likely to have an immediate sensation that your action caused the bang. If you repeatedly feed a goldfish immediately after turning on the light, the fish will soon learn to look for food, in the place you habitually feed it, immediately after the light goes on. Psychologists have developed an extensive system of procedures for testing this kind of learning.

In *classical conditioning,* a stimulus (unconditional stimulus, US) that normally evokes a reflex response (such as the smell of food evoking salivation) is repeatedly presented to an organism, but each time it is preceded by a distinctive stimulus (conditioned stimulus, CS) that does not normally evoke the response (such as a sound, which does not normally evoke salivation). If the conditioning is successful, the organism learns that the sound (CS) is followed by the smell of food (US) and will salivate in response to the sound (CS) when it is presented alone.

Investigators conduct several types of control experiments to ensure that the change in behavior really depends on the association of the two stimuli. For example, there should be no change in response if the sound follows the smell of food, as the sound would then carry no useful information predicting the presence of food. A delay of half a second between the onset of CS and US, or sound and smell, is optimal for most tested mammals. It is also important to demonstrate that when the animal is presented with other types of CS in different experiments, it makes the conditioned response only to the stimulus used in training.

Another popular test of learning is the maze. An animal placed in a maze must choose among two or more alternative pathways at one or more choice points. Rats can learn to navigate through quite complex mazes. For less adept animals, a clear-cut stimulus, such as a lighted pathway versus a dark one, indicates the correct choice, and the maze is simply a practical method of testing an animal's ability to discriminate the stimulus difference and to learn and remember a single correct choice.

In the mid 1960s, Harold J. Morowitz, a well-known biophysicist at Yale University, and his student Philip B. Applewhite began to use microbes to study the nature of memory. They explored learning in a wide variety of microbes and micro-crustaceans. Like others, they observed habituation to mechanical shock, and they noted a common pattern: "The more intense the stimulus, the longer it takes to habituate the organism and the longer the habituation lasts." Applewhite considered there to be extensive evidence that protozoa exhibit habituation. However, in spite of many claims, spread over seven decades, he found no convincing evidence for more sophisticated forms of learning.

When applied to microbes, experiments to demonstrate learning have been plagued by artifacts—many reports of microbes demonstrating associative forms of learning could not be duplicated by other scientists. In many cases, the organisms respond to stimuli that the experimenter does not know about or assumes to be unimportant. An apparatus such as a T-maze may have a slight asymmetry that causes organisms to turn in one direction more than the other, so it is important to test the organisms on several duplicates of the maze or, better, to show that organisms can be trained to turn in either direction. Often what appears to be learning is a change in the environment rather than in the organism. A common problem is that organisms leave chemical stimuli in an apparatus, where they or other individuals will later encounter it. In other cases, electrical shocks or temperature changes produce changes in the chemical composition of the medium, to which the animals later respond. In such cases, naive individuals that are being tested often show the same altered behavior as the "trained" individuals.

Do Flatworms Learn?

The free-living flatworms (Turbellarians) present a very different situation from other microbes. Flatworms are members of the phylum Platyhelminthes, composed of three classes. The trematodes and cestodes are all parasites, but the turbellarians are free-living in aquatic and wet terrestrial habitats. They can be several centimeters long but are flattened, which provides a reduced distance for diffusion to internal tissues. Flatworms move across surfaces by means of a layer of cilia on the under surface (which also secretes mucus) and by waves of muscular contraction. They ingest food into a blind gut, through a mouth located in the middle of the under surface. Because they lack a circulatory system and are otherwise simple in structure, I choose to classify them with microbes.

Flatworms appear to be proficient in associative learning, and because they are the simplest animals with such capabilities, they have been the subjects of many experiments, most employing species of turbellarians known as planaria.

In 1920, P. van Oye obtained evidence that a flatworm could learn to find food that had been placed in a certain location, and his observations were confirmed and amplified in the 1960s by two other research groups independently. In these experiments, food was presented at the end of a wire or glass rod protruding from above into the water in a glass beaker

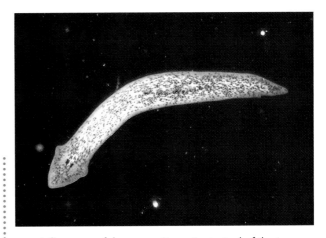

A flatworm of the genus *Dugesia,* typical of the type called planaria, common in freshwater habitats. The pair of dark spots on the head are light receptors.

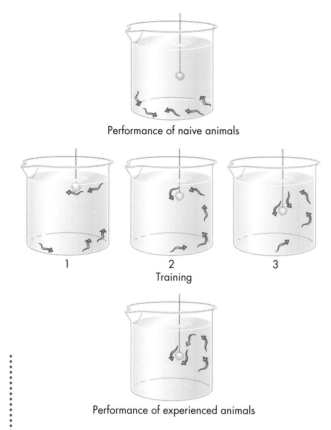

Performance of naive animals

1   2   3

Training

Performance of experienced animals

A common technique for training planaria. Untrained animals rarely find a piece of food that is suspended on a wire or rod away from any of the surfaces on which they crawl, but they can be trained by presenting food on a rod close to the water surface, where they are likely to find it, then gradually lowering the rod so that the food is no longer near the surface.

where the flatworms were living. Untrained flatworms move over the surfaces of the beaker and sometimes across the surface of the water, so at first, the food was held close to the surface, where flatworms find it easily. Subsequently, the rod was lowered, making the food more difficult to locate. Eventually, the flatworms learned to follow the rod down several cen-

timeters below the surface in order to find the food, something that naive animals do not do. If the researchers mixed food scent into the water, trained animals would accumulate at the end of a rod that had no food. If training was conducted in light but the animals were otherwise kept in the dark, then light itself caused the animals to gather at the site at the end of the rod where they had found food before. In one set of six experiments, 0 to 6 percent of untrained planaria reached the end of the rod, while 20 to 40 percent of the trained animals succeeded.

In these experiments, it is not clear exactly what the planaria are learning. They could be learning something complicated like: "When you scent food, move upward if you are on glass, follow the chemical gradient toward metal if you are on the water's surface, and move downward if you are on a wire." Alternatively, they could be learning something simpler like: "If you scent food and encounter a wire, move downward," or even, "If you scent liver, keep moving until you find it."

More clear-cut experiments have been conducted with T-mazes. In these, the animal is placed in the stem and faces a single choice—to move into the left arm or the right. The two arms are distinguished in some way—perhaps one is brightly lit and the other dim. With a modest difference in light intensity, James McConnell at the University of Michigan found no innate preference for light of higher or lower intensity. But when he started rewarding choices, the results changed. If the flatworm moved into the "correct" arm, it was "rewarded" by being returned to its home aquarium for several minutes; animals that moved into the "wrong" arm were "punished" by being returned to the starting position. Animals were subjected to 10 such trials each day for a month. For the first few days, they chose the rewarded arm approximately half the time, as expected if they were making random choices.

By the end of the first week, however, there was a clear preference for the rewarded arm, and by the end of the month the flatworms were moving into the rewarded arm nearly 90 percent of the time.

After leaving the flatworms alone for two or three weeks, the researchers started the training again. The previously trained animals started out choosing the rewarded arm about 80 percent of the time and improved to 90 percent within two weeks. This experiment demonstrates that flatworms can learn to discriminate two arms by their light intensity and that they retain most of the memory for several weeks. A control group was also "punished" at randomly se-

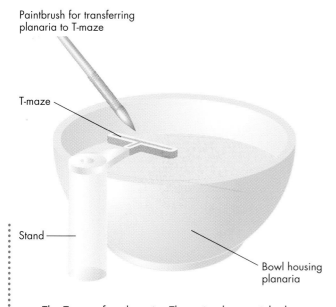

Paintbrush for transferring planaria to T-maze

T-maze

Stand

Bowl housing planaria

The T-maze for planaria. The animals are picked up from the bowl using a paintbrush and placed in a groove in the stem of the T. The animals crawl along the groove and turn right or left at the junction. After they have made a choice, they are picked up and returned to the bowl (as a reward for making a correct choice) or returned to the stem of the T for another chance.

lected times, irrespective of which arm they had moved into, and they continued to choose each arm about 50 percent of the time.

Other investigators have demonstrated that flatworms respond to classical conditioning, one of the standard forms of associative learning. In these experiments, the unconditional stimulus (US) was usually electric shock, which caused the flatworms to contract their body in an avoidance behavior. The conditioned stimulus (CS) was usually light (which can also cause avoidance behavior, most likely head waving). Animals were trained by subjecting them to brief (a few seconds) exposure to the light and, after about a second, a brief exposure to the shock. The researchers selected a second group of animals for a control experiment. The best control experiments still subject the control animals to both stimuli, but simply break the association between the light and the shock so that the light no longer predicts that a shock will occur. The experimental group eventually begins to exhibit avoidance behaviors when exposed to the light more frequently than the control group. The flatworms retained the learned behaviors for more than a month in some experiments.

Planaria have also been conditioned to avoid mechanical vibration, and they can be trained to respond to either light or vibration and not the other. This demonstrates that they learn to associate the specific conditioned stimulus to the unconditioned stimulus and that the behavioral change is part of a general learning mechanism. Flatworms learn to discriminate between light and vibration fairly rapidly compared to some stimuli. Some individuals showed a clear change in behavior after only 15 stimulus pairings.

In the 1960s, over 60 reports on the learning abilities of planaria were published, and about 80 percent of them concluded that planaria had demonstrated some type of associative learning. But some of the most tantalizing experiments of the 1960s, to explore memory transfer, turned out to be a red herring.

Planarians are unusual animals in that they often reproduce by splitting in two: the front half regenerates a new tail and, remarkably, the rear half regenerates a new head. Researchers have also found that they can easily graft parts of one animal onto another animal. These capabilities inspired many scientists to perform regeneration or grafting experiments in the hope of identifying the site of memory.

Some species of planaria are cannibalistic, and in 1962 James McConnell reported that planarians become more sensitive to light after eating animals that had previously been conditioned to avoid light. This observation led to the idea that memories could be transferred from one individual to another by feeding

or injecting the second individual with a chemical extracted from the first. Many researchers hoped that by studying learning transfer in planaria or in mammals they could identify a chemical basis for learning and, in so doing, achieve what could probably be considered the "holy grail" of psychology.

One researcher succeeded in replicating the original transfer of learning experiment in planaria, but none were able to develop reliable protocols that could be used to pin down what chemicals were involved or precisely what information was transferred in planaria or any other animal. As a result of this failure, and a widespread skepticism that chemicals could encode memories, the scientific community lost interest in the possibility of memory transfer and abandoned their studies of planaria. Scientists studying the mecha-

Flatworms Learning to Discriminate

				Number of responses to vibration					
	0	1	2	3	4	5	6	7	8
0				V	VV	V			V
1		L	V	V	V				V
2				V					V
3		L	L	V		V			
4			L	LL	L				
5				L	L		LVL		
6			L		L				
7			L						
8									

Each column corresponds to a certain number of responses to vibration, each row to a certain number of responses to light. Each "V" indicates the response record of an individual trained to respond to vibration, and each "L" the record of an individual trained to respond to light. Of the 25 animals responding unequally to light and vibration, 23 gave more responses to the stimulus they had been trained against. The probability of this many animals favoring the trained response by chance is one in a hundred thousand.

nism of learning have since turned to the fruit fly *Drosophila* (for which genetic dissection techniques are possible) and mollusks (for which microelectrode recording from particular neurons is possible).

It appears that most microbes are capable of only the most specialized forms of learning—sensory adaptation, setting internal clocks, and crude habituation. Many biologists and most psychologists would not consider any of these to be "real" learning. Among animals that could be considered microbes, only the flatworm shows a clear ability for associative learning in a variety of situations. You may wonder why flatworms are so much more capable of learning than nematodes, which also have well-developed nervous systems. One important difference may be that flatworms are much more plastic in their anatomy. Each individual nematode consists of the exact same set of cells (959 somatic cells in the adult hermaphrodite of the nematode *Caenorhabditis elegans,* including 302 neurons) and is incapable of regeneration. This may

have some bearing on the plasticity of connections between neurons, which is probably necessary for learning. Another feature of flatworms that may be important is that they are probably longer lived than nematodes, and learning is more useful to a long-lived animal.

It is clear that learning is not a prominent feature of the behavior of microbes. In spite of the many ingenious mechanisms they have developed to solve their problems, learning is not an important part of their repertoire. It may be that learning is not that useful to such short-lived organisms, and it may be that real learning requires a complex mechanism that is provided only by a complex and malleable nervous system. In fact, generalized associative learning is difficult to demonstrate in any invertebrate, with the exception of the squid and octopus. Insects are certainly capable of learning but only in certain specific circumstances. It appears that effective general mechanisms of learning have only evolved twice—once in the ancestors of the squid and octopus and once in the vertebrates.

Epilogue

As we have seen, microorganisms have evolved a diversity of behaviors to aid in the competition among life forms that is at the foundation of evolution by natural selection. The versatility of these simple organisms may seem surprising at first. But microbes have been evolving five times longer than the arthropods, mollusks, and vertebrates that are more familiar to us, and have had much more time to perfect their behavioral repertoire within the limitations of their small size.

We have also seen that, at this size scale, the laws of physics impose constraints on microbes that are quite different from those that act on large animals like ourselves. Viscosity has much more impact than inertia, and organisms sink slowly in water. More important, in the world of microbes diffusion transports molecules faster than the flow of currents does, although the spread is limited to the vicinity of the source. I have even used this fact to define what a microbe is — an organism sufficiently small that it does not need a circulatory system.

While microbes are inherently fascinating, they also have much to teach us about ourselves. If we look inside our bodies, we find that the constraints imposed by the laws of physics on the scale of microbes are also of great importance to large organisms. Bound by the limitations of diffusion, red blood cells and capillary vessels must be less than 10 μm in diameter, and capillaries must reach within a fraction of a millimeter of tissue everywhere in the body. In carrying out their roles, the individual cells of the body often employ be-

havioral mechanisms similar to those that have proved valuable to microbes. Animal sperm form flagella and swim to the egg, using much the same kind of mechanism that makes flagellated protozoa so mobile. Plant pollen forms tubes that grow to the flower ovary, like fungal hyphae that grow toward a source of nutrients.

Some of the most mobile cells in our bodies are the macrophages that roam in search of infectious bacteria. Their movement resembles that of amoebae, and they are similarly guided by chemical gradients. In the case of macrophages that clear the lungs of inhaled bacteria, Elizabeth Fisher and Douglas Lauffenburger of the University of Pennsylvania have demonstrated that these cells depend on chemotaxis to perform as efficiently as they do. Moving at random, the macrophages could not remove bacteria as fast as necessary.

Molecular biologists have recently discovered that animal cells produce and respond to a wide variety of "growth factors" and similar chemicals of as yet unclear function. Their findings suggest the possibility that additional intriguing behaviors await discovery in the micro world. Researchers are especially puzzled over why many cells both secrete and respond to the same chemicals. Steven Wiley of the University of Utah has proposed that these chemicals serve as a kind of sonar that cells use to assess the local environment. The specific chemicals accumulate nearby, or not, depending on what a cell's neighbors are doing. Its assessment of chemical concentrations thus provides a cell with information that helps it regulate its behavior appropriately.

Through his studies of microbes, Max Delbrück had hoped to discover new laws of physics in the biological world. It appears unlikely, though, that his quest will be successfully concluded even by his successors. Organisms seem to make use of known physical laws, and microbes are no exception—yet the ways they use these laws are so complex, subtle, and varied that investigators will continue to find intriguing new behaviors, and new means of controlling behavior, for a long time to come.

Bibliography

Applewhite, Philip B. 1979. Learning in protozoa. In Biochemistry and Physiology of Protozoa, 2d ed., vol. 1, ed. M. Levandowsky and S. H. Hutner, pp. 341–355. Academic Press, New York.

Aschoff, Jurgen, ed. 1981. Handbook of Behavioral Neurobiology. Vol. 4, Biological Rhythms. Plenum Press, New York.

Barron, G. L. 1977. The Nematode-Destroying Fungi. Canadian Biological Publications Ltd., Guelph, Ontario.

Barron, G. L. 1990. A new predatory Hypomycete capturing copepods. Can. J. Bot. 68:691–696.

Bellows, A., H. G. Trüper, M. Dworkin, W. Harder, and K.-H. Schleifer. 1992. The Prokaryotes, 2d ed. Springer-Verlag, New York.

Berg, H. C. 1983. Random Walks in Biology. Princeton University Press, Princeton.

Buskey, E. J., and E. Swift. 1983. Behavioral responses of the coastal copepod Acartia husonica (Pinhey) to simulated dinoflagellate bioluminescence. J. Exp. Mar. Biol. Ecol. 72:43–58.

Cerdá-Olmedo, E., and E. D. Lipson. 1987. Phycomyces. Cold Spring Harbor Laboratory, Cold Spring Harbor, New York.

Charmichael, W. W. 1994. The toxins of cyanobacteria. Scientific American, January, pp. 78–86.

Corning, W. C., and S. Kelly. 1973. Platyhelminthes: the turbellarians. Chap. 4 in Invertebrate Learning, vol. 1, Protozoans Through Annelids, ed. W. C. Corning, J. A. Dyal, and A. O. D. Willows, pp. 171–224. Plenum Press, New York.

Croll, N. A. 1970. The Behaviour of Nematodes: St. Martin's Press, New York.

Davidson, N. 1962. Statistical Mechanics. McGraw-Hill, New York.

Donchaster, C. C. 1971. Feeding in plant parasitic nematodes: mechanism and behavior. Chap. 19 in Plant Parasitic Nematodes, ed. B. M. Zuckerman, W. F. Mai, and R. A. Rhode, pp. 137–157. Academic Press, New York.

Doughty, M. J., and S. Dryl. 1981. Control of ciliary activity in Paramecium: an analysis of chemosensory transduction in a eukaryotic unicellular organism. Progress in Neurobiology 16:1–115.

Douglas, A. E. 1994. Symbiotic Interactions. Oxford University Press, New York.

Dusenbery, D. B. 1980. Responses of the nematode Caenorhabditis elegans to controlled chemical stimulation. J. Comp. Physiol. 136:327–331.

Dusenbery, D. B. 1989. The value of asymmetric signal processing in klinokinesis. Biological Cybernetics 61:401–404.

Dusenbery, D. B. 1989. A simple animal can use a complex stimulus pattern to find a location: nematode thermotaxis in soil. Biological Cybernetics 60:431–437.

Dusenbery, D. B. 1992. Sensory Ecology. W. H. Freeman, New York.

Dusenbery, D. B., and T. W. Snell. 1995. A critical body size for use of pheromones in mate location. Journal of Chemical Ecology 21:427–438.

Edmunds, Leland N. 1988. Cellular and Molecular Basis of Biological Clocks. Springer-Verlag, New York.

Foster, K. W., and R. D. Smyth. 1980. Light antennas in phototactic algae. Microbial Reviews 44:572–630.

Gregory, P. H., Guthrie, E. J., and Bunce, M. E. 1959. Journal of General Microbiology 20:328–354.

Ingold, C. T. 1965. Spore Liberation. Clarendon Press, Oxford.

Jennings, H. S. 1904. Contributions to the Study of the Behavior of the Lower Organisms. Carnegie Institution of Washington, D.C.

Bibliography

Larkin-Thomas, Patricia L., Gary B. Coté, and Stuart Brody. 1990. Circadian rhythms in *Neurospora crassa:* biochemistry and genetics. Critical Reviews in Microbiology 17:365–416.

Macfarlane, G. 1984. Alexander Fleming. The Hogarth Press, London.

Maggenti, A. 1981. General Nematology. Springer-Verlag, New York.

Melkonian, M. 1992. Algal Cell Motility. Chapman and Hall, New York.

Page, R. M. 1964. Science 146:925–927.

Pelczar, M. J., Jr., E. C. S. Chan, and N. R. Krieg. 1993. Microbiology: Concepts and Applications. McGraw-Hill, New York.

Poinar, George O., Jr. 1983. The Natural History of Nematodes. Prentice Hall, Englewood Cliffs, New Jersey.

Prescott, L. M., J. P. Harley, and D. A. Klein. 1990. Microbiology. Wm. C. Brown Publishers, Dubuque, Iowa.

Roberts, A. M. 1981. Hydrodynamics of protozoan swimming. In Biochemistry and Physiology of Protozoa, vol. 4, ed. M. Levandowsky and S. H. Hutner, pp. 5–66. Academic Press, New York.

Stephens, K. 1986. Pheromones among the procaryotes. Critical Reviews in Microbiology 13:309–334.

Walsby, A. E. 1978. The gas vesicles of aquatic prokaryotes. Symp. Soc. gen. Microbiol. 28:327–358.

Winfree, Arthur T. 1987. The Timing of Biological Clocks. Scientific American Library, New York.

Sources of Illustrations

CHAPTER ONE

Facing page 1: Barry J. Wicklow, Biology Department, Saint Anselm College

Page 3: William Heath, 1828. Philadelphia Museum of Art. Gift of Mrs. William H. Horstmann

Page 4: Barry Dowsett/Science Photo Library/Photo Researchers

Page 9: M. I. Walker/Natural History Photo Agency

Page 10: Tony Brain/SPL/Photo Researchers

Page 11: C. R. Woese. 1994. *Microbiological Reviews* 58:1–9, fig. 1

Page 13: J. C. Stevenson/Animals Animals, Earth Scenes

Page 15: Holger W. Jannasch, Woods Hole Oceanographic Institution

Page 16: Jeremy Pickett-Heaps, University of Melbourne

CHAPTER TWO

Page 18: Karl Aufderheide/Visuals Unlimited

Page 22: S. Vogel. 1981. *Life in Moving Fluids.* Princeton University Press, Princeton, fig. 5.5

Page 23: D. J. Patterson/Planet Earth Pictures

Page 26: C. J. Jones and S.-I. Aizawa. 1991. *Advances in Microbial Physiology* 32:109–171

Page 28 (bottom left): Fred Hossler/Visuals Unlimited

Page 28 (top right): Eric Gravé/Photo Researchers

Page 29: H. C. Berg. 1976. *J. Theor. Biol.* 56: 269–273

Page 30: R. Eckert. 1988. *Animal Physiology,* 3rd ed. W. H. Freeman and Company, New York

Page 31 (left): M. I. Walker/Photo Researchers

Page 31 (right): U. Ruffer and W. Nultsch. 1985. *Cell Motility* 5:251–263

Page 32: Biophotos/Photo Reseachers

Page 33: M. Abbey/Visuals Unlimited

Page 34: I. Gibbons/Photo Researchers

Page 35: R. Eckert. 1988. *Animal Physiology,* 3rd ed. W. H. Freeman and Company, New York

Page 37: Sinclair Stammers/SPL/Photo Researchers

Page 38: Patrice Albert

Page 39: E. M. Reed and H. R. Wallace. 1965. *Nature* 206:210–211

Pages 40 and 41: H. S. Jennings. 1904. *Contributions to the Study of the Behavior of the Lower Organisms.* Carnegie Institution of Washington

Page 42: D. Porter

Page 43: A. S. Edwards, Planet Earth Pictures

Page 44: L. A. Edgar and J. D. Pickett-Heaps. 1984. *Progress in Phycological Research,* vol. 3, ed. F. E. Round and David J. Chapman. Biopress Ltd., Bristol

CHAPTER THREE

Page 46: Hans Reichenbach, GBF, Braunschweig

Page 48 (left): G. I. Bernard/Natural History Photo Agency

Page 48 (right): Jeremy Burgess/SPL/Photo Researchers

Page 49: George L. Barron, University of Guelph

Page 50: George L. Barron, University of Guelph

Page 52 (right): C. T. Ingold. 1965. *Spore Liberation.* Clarendon Press, Oxford

Page 52 (left): S. Nielsen/DRK

Page 53: D. S. Meredith. 1961. *Annals of Botany, N.S.* 25(99):271–278

Page 54: Michael Fogden, DRK

Page 55: Matt Meadows/Peter Arnold, Inc.

Page 56: D. B. Dusenbery

Page 57: George L. Barron, University of Guelph

Page 59: C. T. Ingold. 1965. *Spore Liberation.* Clarendon Press, Oxford

Page 60: Matt Meadows/Peter Arnold, Inc.

Page 61 (bottom left): Veronika Burmeister/Visuals Unlimited

Page 61 (top right): David M. Phillips/Visuals Unlimited

CHAPTER FOUR

Page 62: Pseudocolor image of Dictyostelium aggregation provided by Dr. P. R. Fisher, School of Microbiology, La Trobe University, Melbourne, Australia. (World Wide Web Page: http://luff.latrobe.edu.au/~micprf/my.html)

Page 66: G. L. Hazelbauer. 1988. The bacterial chemosensory system. *Can. J. Microbiol.* 34:466–474

Page 69: J. E. Segall, S. M. Block, and H. C. Berg. 1986. Temporal comparisons in bacterial chemotaxis. *Proceedings of the National Academy of Sciences, U.S.A.* 83:8987–8991

Page 70: D. B. Dusenbery. 1980. Responses of the nematode *Caenorhabditis elegans* to controlled chemical stimulation. *Journal of Comparative Physiology* 136:327–331

Page 71: D. B. Dusenbery. 1989. The value of asymmetric signal processing in klinokinesis. *Biol. Cybernetics* 61:401–404

Page 73: Susan Barns, Indiana University

Page 74: M. J. Grimson and R. L. Blanton, Texas Tech University

Page 75: J. E. Segall and G. Gerisch. 1989. Genetic approaches to cytoskeleton function and the control of cell motility. *Current Opinion in Cell Biology: Cytoplasm and Cell Motility* 1:44–50

Page 77: D. B. Dusenbery

Pages 78 and 79: K. W. Foster and R. D. Smyth. 1980. *Microbiological Reviews* 44: 572–630

Page 80: Kenneth Foster, Syracuse University

Page 81: Kim Taylor/Bruce Coleman, Inc.

Page 82: John Hodgin

Page 84: H. Machemer and J. W. Deitmer. 1985. *Progress in Sensory Physiology,* vol. 5. Springer-Verlag, New York

Page 86: David Dennison, Dartmouth College

Page 87: David Dennison, Dartmouth College

Page 88: W. Shropshire, Jr. and J.-F. Lafay. 1987. Sporangiophore and mycelial responses to stimuli other than light. In *Phycomyces,* ed. E. Cerdá-Olmedo and E. D. Lipson, pp. 127–154. Cold Spring Harbor Laboratory, Cold Spring Harbor, New York

CHAPTER FIVE

Page 90: Kim Taylor/Bruce Coleman, Inc.

Page 92: Robert Arnold/Planet Earth Pictures

Page 93: Daniel Branton, Harvard University

Page 94: F. Garcia-Pitchel, M. Mechling, and R. Castenholz. 1994. *Applied Environmental Microbiology* 60:1500–1511

Page 96 (left): E. R. Degginger/Bruce Coleman

Page 96 (right): M. Z. Gliwicz. 1986. *Nature* 320:746–748

Page 98 (left): Eric V. Gravé/Photo Researchers

Page 98 (right): Jeremy Pickett-Heaps, University of Melbourne

Page 99: T. Fenchel and B. J. Finlay. 1986. *J. Protozool.* 33(1):69–76

Page 101: D. Balkwill and D. Maratea/Visuals Unlimited

Page 102: Kent Wood/Peter Arnold, Inc.

Page 103: J. A. Diez and D. B. Dusenbery. 1989. Preferred temperature of *Meloidogyne incognita. Journal of Nematology* 21(1):99–104

Page 105: D. B. Dusenbery. 1988. Behavioral responses of *Meloidogyne incognita* to small temperature changes. *Journal of Nematology* 20:351–355

Page 106: Computer image by D. B. Dusenbery

Page 107: Computer image by D. B. Dusenbery

Page 108: D. B. Dusenbery. 1988. Avoided temperature leads to the surface: computer modeling of slime mold and nematode thermotaxis. *Behav. Ecol. Sociobiol.* 22:219–223

Page 109: Computer image by D. B. Dusenbery

CHAPTER SIX

Page 110: George L. Barron, University of Guelph

Page 112: George L. Barron, University of Guelph

Page 114: George L. Barron, University of Guelph

Page 115: George L. Barron, University of Guelph

Pages 116 and 117: G. L. Barron. 1977. *The Nematode-Destroying Fungi. Topics in Mycology,* no. 1. Canadian Biological Publications

Page 118: George L. Barron, University of Guelph

Page 119: Michael Abbey

Page 121 (left): Jeremy Burgess/SPL/Photo Researchers

Page 121 (right): Jeremy Burgess/SPL/Photo Researchers

Page 122: Manfred Kage/Peter Arnold Inc.

Page 123: D. Cavagnaro/DRK

Page 126 (left): K. H. Nealson and J. W. Hastings. 1992. In *The Prokaryotes: A Handbook on the Biology of Bacteria: Eco-*

physiology, Isolation, Identification, Applications, ed. Albert Balows. Springer-Verlag, New York

Page 126 (right): Photo courtesy of George O. Poinar

Page 127: C. R. Calention/Visuals Unlimited

CHAPTER SEVEN

Page 132: Helga Lade/Peter Arnold, Inc.

Page 134 (left): Biophoto Associates/Photo Researchers

Page 134 (right): Gary W. Grimes and S. W. L'Hernault

Page 135: A. S. Dabholkar, Wright State University

Page 136: Wayne W. Carmichael. January 1994. The toxins of cyanobacteria. *Scientific American* 270(1):78–86

Page 137: Pete Atkinson/Planet Earth Pictures

Page 138: A. M. Kelly, C. C. Kohler, and D. R. Tindall. 1992. *Environmental Biology of Fishes* 33:275–286

Page 139: Mark Mattock/Planet Earth Pictures

Page 141 (top): Jürgen Kusch, Westfälische Wilhelms-Universität

Page 141 (bottom): J. Kusch. 1993. *Oecologia* 94(4): 571–575

Page 142: Barry J. Wicklow, Biology Department, Saint Anselm College

Page 144 (left): P. R. Johnsson and P. Tiselius. 1990. *Marine Ecology Progress Series* 60:35–44

Page 144 (right): R. L. Wallace and T. W. Snell. 1991. Rotifera. In *Ecology and Systematics of North American Freshwater Invertebrates,* ed. J. H. Thorp and A. P. Covich, pp. 187–248. Academic Press, San Diego

Page 145: D. J. Hibberd. 1970. *Br. Phycol. J.* 5(2):119–143

Page 146 (left): Manfred Kage/Peter Arnold, Inc.

Page 146 (right): E. J. Buskey, L. Mills, and E. Swift. 1983. *Limnol. Oceanog.* 28(3): 575–579

Page 147: D. B. Dusenbery

Page 150: M. L. Lemos, C. P. Dopazo, A. E. Toranzo, and J. L. Barja. 1991. *J. Applied Bacteriology* 71:228–232

Sources of Illustrations

Page 152: St. Mary's Hospital Medical School/SPL/Photo Researchers

Page 153: Biophoto Associates/Photo Researchers

Page 156: M. Mazzola et al. 1992. *Applied Environmental Microbiology* 58(8):2616–2624

Page 157: D. S. Messina and A. L. Baker. 1982. *J. Plankton Research* 4(1):41–46

CHAPTER EIGHT

Page 158: Clay H. Wiseman/Animals Animals, Earth Scenes

Page 160: J. W. Hastings and J. G. Morin. Bioluminescence. 1991. In *Neural and Integrative Animal Physiology,* ed. C. L. Prosser, pp. 131–170. Wiley-Liss, New York

Page 161: James G. Morin/Visuals Unlimited

Page 165: Bruce Iverson

Page 166: Jeremy Pickett-Heaps, University of Melbourne

Page 167 (left): Giulio Melone, Department of Biology, University of Milan

Page 167 (right): T. W. Snell, R. Rico-Martinez, L. N. Kelly, and T. E. Battle. 1995. *Marine Biology* 123:347–353

Page 169 (left): Reinhard Wirth, Universität Regensburg

Page 169 (right): D. B. Clewell. 1993. Cell 73(1):9–12

Page 172: Dwight Kuhn

Page 173: J. R. Raper. 1970. Chemical ecology among lower plants. In *Chemical Ecology,* ed. E. Sonheimer and J. B. Simeone, pp. 21–42. Academic Press, San Diego

Page 175: Peter Devreotes, Department of Biological Chemistry, Johns Hopkins University School of Medicine

CHAPTER NINE

Page 176: J. Woodland Hastings, Harvard University

Page 178: J. Woodland Hastings, Harvard University

Page 179: J. W. Hastings, B. Rusak, and Z. Boulos. 1991. In *Neural and Integrative Animal Physiology,* ed. C. L. Prosser, pp. 435–546. Wiley-Liss, New York

Page 180: T. Roennenberg and D. Morse. 1993. *Nature* 362:362–364

Page 183: A. Mitsui, S. Kumazawa, A. Takahashi, H. Ikemoto, S. Cao, and T. Arai. *Nature* 323:720–722

Page 184: F. E. Round. 1984. *The Ecology of Algae.* Cambridge University Press, Cambridge

Page 185: J. D. Palmer and F. E. Round. 1967. *Biological Bulletin* 132:44–55

Page 186: J. W. Hastings, B. Rusak, and Z. Boulos. 1991. In *Neural and Integrative Animal Physiology,* ed. C. L. Prosser, pp. 435–546. Wiley-Liss, New York

CHAPTER TEN

Page 188: Manfred Kage/Peter Arnold, Inc.

Page 190: E. R. Degginger/Animals Animals, Earth Scenes

Page 191: D. J. Patterson. 1973. *Behaviour* 45:304–311

Page 195: Dwight Kuhn

Page 196: W. C. Corning and S. Kelly. 1973. In *Invertebrate Learning,* vol. 1, ed. W. C. Corning, J. A. Dyal, and A. O. D. Willows. Plenum Press, New York

Page 197: J. V. McConnell. 1967. In *Chemistry of Learning,* ed. W. C. Corning and S. C. Ratner, pp. 217–233. Plenum Press, New York

Index

Note: Page references to photographs are in italic type. The letter "t" after a page reference indicates a table.

Index

Selected Books in the Scientific American Library Series

This book is due for return on or before the last date shown below.

9 - MAR 2011